Reviews of Physiology, Biochemistry and Pharmacology 153

Reviews of

153 Physiology Biochemistry and Pharmacology

Editors
S.G. Amara, Pittsburgh • E. Bamberg, Frankfurt
R. Jahn, Göttingen • W.J. Lederer, Baltimore
A. Miyajima, Tokyo • H. Murer, Zürich
S. Offermanns, Heidelberg • G. Schultz, Berlin
M. Schweiger, Berlin

With 9 Figures in color

Springer

Library of Congress-Catalog-Card Number 74-3674

ISSN 0303-4240
ISBN 3-540-24012-8 Springer Berlin Heidelberg New York

Springer is a part of Springer Science+Business Media

springeronline.com

Printed in The Netherlands

Editor: Dr. Thomas Mager, Heidelberg
Desk editor: Anne Clauss, Heidelberg
Production editor: Andreas Gösling, Heidelberg
Cover design: design & production GmbH, Heidelberg
Typesetting: Stürtz GmbH, Würzburg

Printed on acid-free paper – 14/3150 ag 5 4 3 2 1 0

Rev Physiol Biochem Pharmacol (2005) 153:1–46
DOI 10.1007/s10254-004-0025-5

M. P. Mayer

Recruitment of Hsp70 chaperones: a crucial part of viral survival strategies

Published online: 9 July 2004

Abstract Virus proliferation depends on the successful recruitment of host cellular components for their own replication, protein synthesis, and virion assembly. In the course of virus particle production a large number of proteins are synthesized in a relatively short time, whereby protein folding can become a limiting step. Most viruses therefore need cellular chaperones during their life cycle. In addition to their own protein folding problems viruses need to interfere with cellular processes such as signal transduction, cell cycle regulation and induction of apoptosis in order to create a favorable environment for their proliferation and to avoid premature cell death. Chaperones are involved in the control of these cellular processes and some viruses reprogram their host cell by interacting with them. Hsp70 chaperones, as central components of the cellular chaperone network, are frequently recruited by viruses. This review focuses on the function of Hsp70 chaperones at the different stages of the viral life cycle emphasizing mechanistic aspects.

Introduction

The life cycle of a virus is a course with many obstacles that must be overcome in order to produce a sufficient number of progeny to guarantee evolutionary survival (Fig. 1). Viruses have to interact with cell surface receptors, induce endocytosis and/or membrane fusion and thereby achieve entry into the cell. Their capsid, which needs to be relatively stable outside of the cell to ensure sufficient protection of the viral genome against environmental impacts, must be disassembled or opened to allow the viral nucleic acid to gain access to the cytoplasm and nucleus where replication and transcription take place. The maturation process of viral proteins, which often consist of multiple domains or are produced as polyproteins, can be very complicated. New capsids have to assemble in an ordered way around the viral genome and release of new virions must be induced. In addition, cellular

M. P. Mayer (✉)
Zentrum für Molekulare Biologie, Universität Heidelberg, Im Neuenheimer Feld 282, 69120 Heidelberg, Germany
e-mail: M.Mayer@zmbh.uni-heidelberg.de · Tel.: +49-6221-546829 · Fax: +49-6221-545894

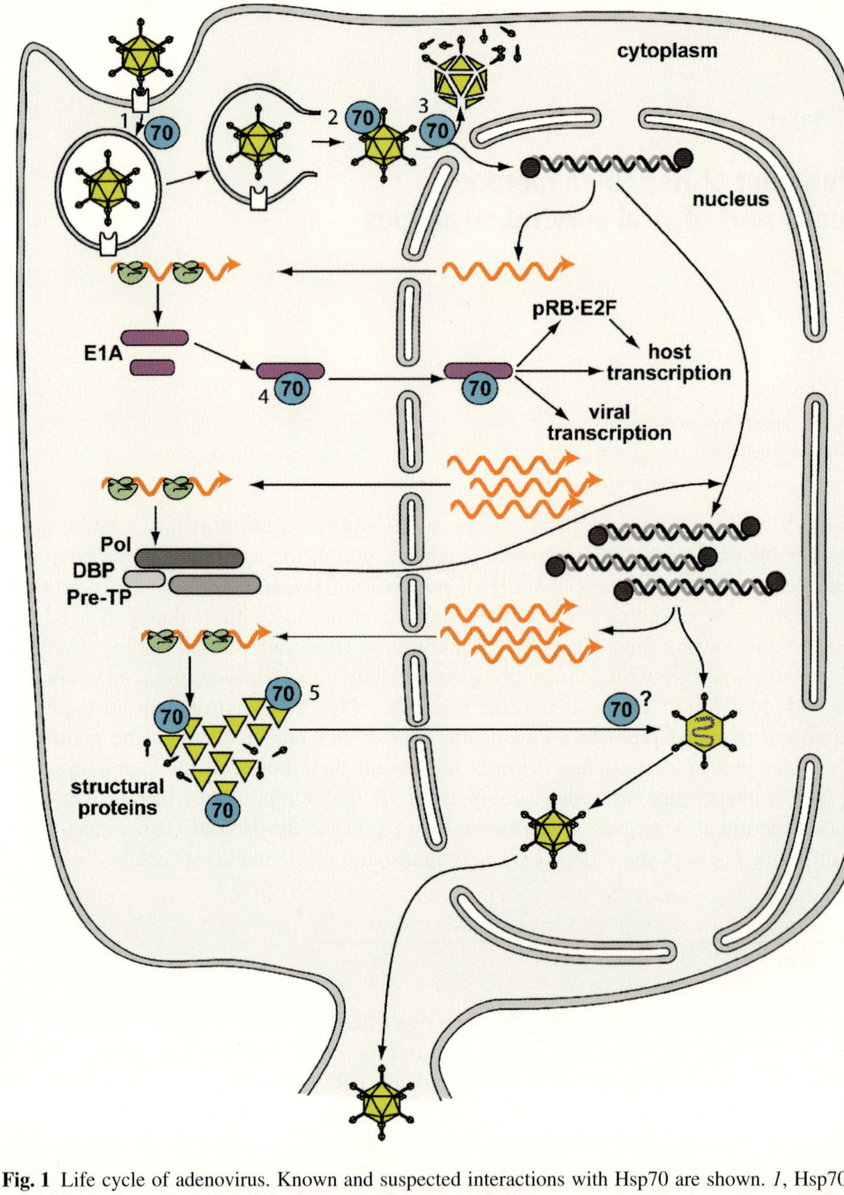

Fig. 1 Life cycle of adenovirus. Known and suspected interactions with Hsp70 are shown. *1*, Hsp70 is involved in the recycling of clathrin during the formation of clathrin coated pits and vesicles and afterwards in the uncoating of clathrin coated vesicles (Greene and Eisenberg 1990; Newmyer et al. 2003; Newmyer and Schmid 2001; Ungewickell 1985). *2*, Binding of Hsp70 to the hexon capsid protein. *3*, Hsp70-mediated release of the viral genome into the nucleus. *4*, Interaction of Hsp70 with the immediate early gene product E1A, dissociation of pRB·E2F complexes, regulation of viral and host transcription by E1A with likely participation of Hsp70. *5*, interaction of Hsp70 with newly synthesized coat proteins and possible role in the virion assembly

defense mechanisms must be overcome and sometimes cell differentiation prevented and start of the cell cycle induced.

Some of these obstacles involve protein folding processes and it is therefore not surprising that most viruses interact with cellular chaperones. In fact, two of the major chaperone systems in *Escherichia coli*, the Hsp70 (DnaK, DnaJ, GrpE) and the Hsp60 (GroEL, GroES) systems, were originally discovered as host factors essential for growth of bacterial viruses, the bacteriophages λ and T4 (Georgopoulos 1972, 1977; Sunshine et al. 1977). Early on it became clear that the folding tasks of the chaperones, in particular of the Hsp70s, involves not only the acceleration of the maturation of viral proteins but also the regulation of the viral life cycle and coordination of host and viral physiological states. In eukaryotic cells Hsp70 chaperones are involved in the regulation of fundamental cellular processes such as the cell cycle and apoptosis. The functional interaction of viruses with these chaperones therefore contributes to reprogramming the host cell, specifically to allow re-entry into the cell cycle and to avoid premature apoptosis. Hsp70 chaperones also seem to be involved in circumvention of cellular defense and sometimes even in avoidance of the host defense mechanisms.

This treatise deals mainly with the role of Hsp70 chaperones for virus proliferation. Before detailing the virus–Hsp70 interactions, the cellular functions and the molecular mechanism of Hsp70 chaperones and their various co-chaperones are introduced. The role of Hsp70 at different stages of the viral life cycle is discussed. Finally, an evolutionary facet of the virus–Hsp70 relationship will be considered.

Mechanism of Hsp70 chaperones

Members of the Hsp70 family of chaperones are involved in an astonishingly large variety of processes. Among these processes are the folding of newly synthesized polypeptides, the refolding of stress denatured proteins, the disaggregation of protein aggregates, the translocation of organellar and secretory proteins across membranes, the assembly and disassembly of oligomeric structures, and the control of the biological activity and stability of regulatory proteins (Bukau et al. 2000; Craig et al. 1999; Gething 1999; Hartl and Hayer-Hartl 2002; Neupert and Brunner 2002; Ryan and Pfanner 2002; Schlieker et al. 2002). Hsp70 chaperones not only continuously survey the folding status of proteins as part of their quality control function that is especially important under stress conditions, they are also involved in many cellular housekeeping functions including signal transduction and regulation of cell cycle and cell death (Beere and Green 2001; Helmbrecht and Rensing 1999). Among these housekeeping functions, it is especially noteworthy that in most organisms Hsp70s are involved in the regulation of the stress response (Arsene et al. 2000; Gabai et al. 1998; Morimoto 1999; Urano et al. 2000). In eukaryotic cells Hsp70 chaperones are found in virtually all compartments and even on the cell surface where specific receptors exist for the binding of Hsp70 proteins (Asea 2003).

The evolutionary adaptation to such a broad spectrum of functions was made possible by three basic properties of Hsp70s. First, they transiently interact with short hydrophobic peptide stretches within their target proteins and protein size is therefore not a limiting factor. Second, they are regulated in their activity by co-chaperones including the large family of modular J-domain proteins (JDPs) that target Hsp70s to their substrates. Third, for specific tasks they cooperate with other chaperone systems.

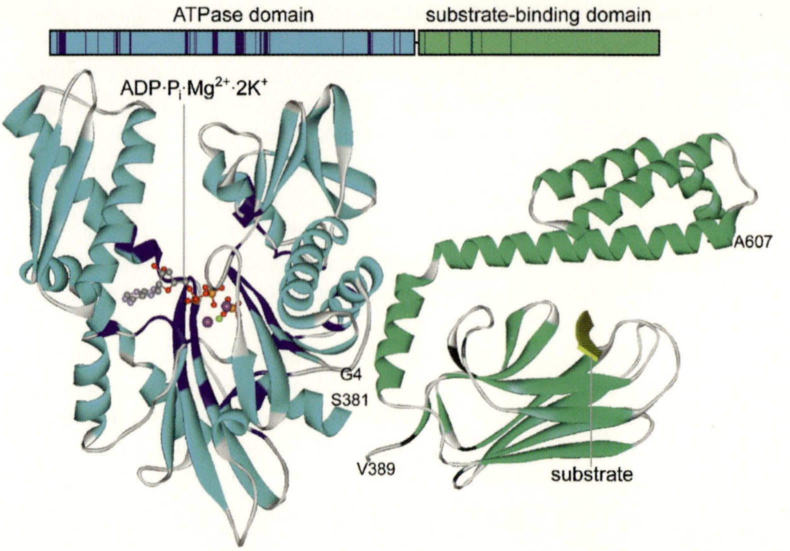

Fig. 2 Structure of Hsp70 chaperones. *Top*: domain structure of Hsp70s; the residues that are conserved in at least 11 out of the 12 known full-length sequences of viral Hsp70s are shown in *darker colors*. *Bottom*: secondary structure representation of the crystal structure of the ATPase domain of bovine Hsc70 (1BUP; *left*; Sousa and McKay 1998) and the substrate-binding domain of *E. coli* DnaK (1DKX; *right*; Zhu et al. 1996). In viral Hsp70s conserved residues are shown in *dark blue* (ATPase domain) and *dark green* (substrate-binding domain)

The ATPase cycle

Hsp70 homologs share the same overall structure, consisting of an N-terminal ATPase domain of 45 kDa and a C-terminal substrate binding domain of at least 25 kDa which is further subdivided into a β-sandwich subdomain of 15 kDa and a C-terminal α-helical subdomain (Fig. 2). ATP binding to the ATPase domain of Hsp70 proteins decreases the affinity of the substrate-binding domain for substrates by 5- to 85-fold (Mayer et al. 2000b; Palleros et al. 1993; Schmid et al. 1994). This decrease in affinity is due to an increase in the dissociation rate (k_{off}) of Hsp70-substrate complexes by two to three orders of magnitude and a concomitant increase of the association rate (k_{on}) for substrate binding by approximately 50-fold (Mayer et al. 2000b; Pierpaoli et al. 1997; Schmid et al. 1994; Theyssen et al. 1996). The ATPase cycle of Hsp70 thus consists of an alternation between the ATP state with low affinity and fast exchange rates for substrates, and the ADP state with high affinity and low exchange rates for substrates.

ATP hydrolysis by Hsp70s is generally very slow (τ=5–15 min) but is stimulated by substrate association (two- to tenfold) and by a J-domain containing co-chaperones (in general $\leq$tenfold). The simultaneous interaction of Hsp70s with a substrate and a JDP synergistically stimulate the ATPase activity up to several thousand-fold (Barouch et al. 1997; Karzai and McMacken 1996; Laufen et al. 1999; Misselwitz et al. 1998). After ATP hydrolysis the substrate is tightly bound by the Hsp70 chaperone and for most Hsp70s nucleotide exchange is rate-limiting for substrate release, i.e., the rate with which ADP dissociates determines how long the substrate remains in complex with the Hsp70 proteins. For some prokaryotic, mitochondrial and plastidal Hsp70s nucleotide exchange is cat-

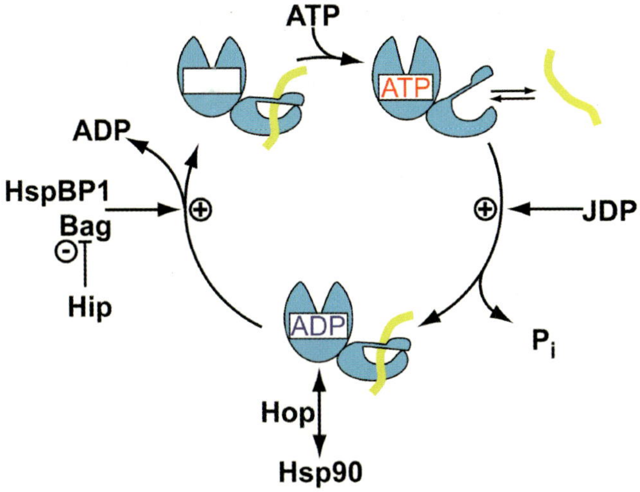

Fig. 3 ATPase cycle of Hsp70 chaperones and action of some co-chaperones on the ATPase cycle

alyzed by the nucleotide exchange factor GrpE (Harrison et al. 1997; Liberek et al. 1991; Packschies et al. 1997). For the cytosolic Hsc70-type Hsp70s the family of Bag proteins has been shown to accelerate ADP dissociation (Brehmer et al. 2001; Gässler et al. 2001; Höhfeld and Jentsch 1997; Sondermann et al. 2001) (Fig. 3).

Hsp70 substrate interactions

Hsp70 chaperones interact promiscuously with virtually all unfolded proteins but generally do not bind their native counterparts. Yet, they also recognize certain folded proteins with high specificity. An important question therefore is, how Hsp70 can combine within its substrate specificity both of these seemingly contradicting properties. Using a library of cellulose-bound peptides scanning the sequences of natural proteins the binding motif for the *E. coli* homolog DnaK was elucidated (Rüdiger et al. 1997). This motif consists of a core of five amino acids enriched in hydrophobic residues, flanked on both sides by a region where positively charged residues are preferred. Such motifs occur in virtually all proteins on average every 30–40 residues. In folded proteins they are mostly found in the hydrophobic core explaining the promiscuous binding to denatured proteins. In contrast, it is not completely clear how Hsp70s recognize specifically certain folded proteins, although the binding sites of Hsp70 proteins in a few native substrates has been determined (Hoff et al. 2002; Kim et al. 2002; M.P. Mayer and B. Bukau, unpublished results).

How Hsp70 systems refold denatured proteins is also still enigmatic. In analogy to the global unfolding hypothesis proposed for Hsp60 and Hsp100 chaperones (Shtilerman et al. 1999; Weber-Ban et al. 1999) it is proposed that Hsp70s induce local conformational changes in their substrate protein thereby giving them a new chance to fold productively (Mayer et al. 2000a; Pierpaoli et al. 1997; Slepenkov and Witt 2002).

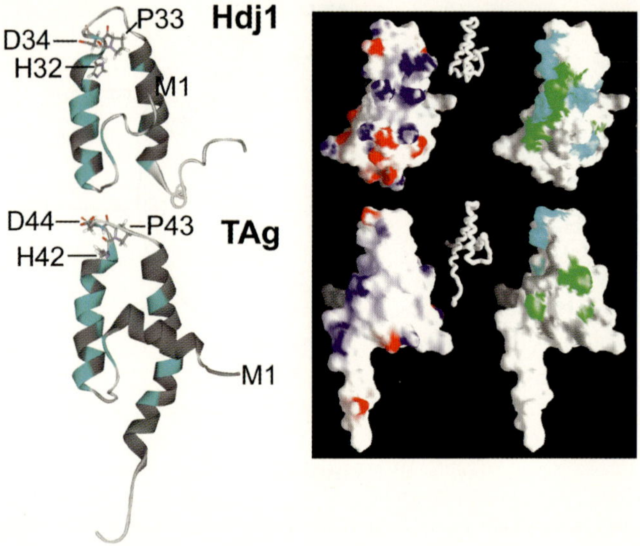

Fig. 4 Comparison of the J domains of Hdj1 and polyomavirus T antigen. *Left*: Secondary structure representation of one of the 20 energy minimized NMR structures of human Hdj1 (1HDJ; Qian et al. 1996) and mouse polyomavirus T antigen (1FAF; Berjanskii et al. 2000) with the highly conserved and essential His–Pro–Asp motif as stick model. The positions of the residues are shown in *cyan* for which in the corresponding residues in *E. coli* DnaJ J-domain line broadening and chemical shift perturbation is observed upon interaction with the *E. coli* Hsp70 homolog DnaK ATPase domain. *Right*: Surface representations of the two J-domains colored according to the surface potential (–10 to +10; *left*) and the residues that correspond to the DnaK interacting residues in DnaJ (*right*); *cyan*, identical residues as in DnaJ; *green*, conservative exchanges; *gray*, not conserved. The *inset* shows the orientation of the J-domains as worm representation. The surface representation and the electrochemical potentials were calculated using the GRASP program

The family of J-domain proteins: targeting of substrates to Hsp70s

The family of JDPs consists of modular multidomain proteins that are characterized by a conserved domain of 70–80 amino acids, the so-called J-domain (Cheetham and Caplan 1998; Kelley 1998; Laufen et al. 1998) (Fig. 4). The additional domains of JDPs serve as protein–protein interaction sites allowing JDPs to bind substrate proteins, to interact with other chaperones, or to target JDPs to specific cellular locations.

The J-domain is essential for the functional interaction of JDPs with Hsp70s, i.e., the stimulation of the ATPase activity, and mutations within this domain especially in the almost universally conserved tripeptide motif His-Pro-Asp (HPD-motif) abrogate the function of the JDPs as co-chaperones of Hsp70s. JDPs have been shown to promote the binding of Hsp70s to their substrates by simultaneous co-stimulation of the ATPase activity leading to the locking-in of the substrate into the substrate binding cavity of the Hsp70s (Karzai and McMacken 1996; Laufen et al. 1999; Misselwitz et al. 1998). This function of JDPs depends on close proximity of JDP and substrate, which for most JDPs is guaranteed by direct interaction with the Hsp70 substrate. How JDPs interact with substrates and mediate their transfer onto Hsp70 partner proteins is not clear.

Some JDPs have a broad substrate specificity such as *E. coli* DnaJ, yeast Ydj1 and human DjB1/Hdj1 and are able to prevent the aggregation of misfolded proteins, while others have more restricted substrate spectra, such as the clathrin-specific auxilin, or they may not bind substrates themselves but rather be positioned in close proximity of substrates. The latter seems to be the case for Sec63 at the translocation pore in the endoplasmic reticulum (ER) (Corsi and Schekman 1996; Rapoport et al. 1996), Pam18 at the translocase of the inner mitochondrial membrane (D'Silva et al. 2003; Mokranjac et al. 2003; Truscott et al. 2003) and cysteine string proteins on the surface of neurosecretory vesicles (Buchner and Gundersen 1997; Evans et al. 2003).

GrpE and the Bag family of proteins: induction of substrate unloading

The 21-kDa nucleotide exchange factor GrpE has been found only in prokaryotic organisms and in some eukaryotic organelles (mitochondria and plastids). It forms an asymmetric homodimer when in complex with its Hsp70 partner protein DnaK and does not share any sequence or structural motifs with other known proteins. The family of Bag proteins are multidomain proteins that share a three-helix bundle domain of approximately 120 residues, the so-called Bag domain, which is essential for interaction with Hsp70 proteins (Briknarova et al. 2001; Sondermann et al. 2001; Takayama et al. 1999).

Bag proteins differ from GrpE proteins by their ability to associate with ligands other than Hsp70 proteins. These interactions link Bag proteins to a number of highly diverse cellular processes including signal transduction processes inducing apoptosis, proliferation, and differentiation. Bag proteins have been found in complexes with the anti-apoptotic Bcl-2, the protein kinase Raf, the transcription factor c-Jun, the receptors for vitamin D, androgen and glucocorticoids (Bardelli et al. 1996; Clevenger et al. 1997; Schneikert et al. 1999; Takayama et al. 1995; Wang et al. 1996; Zeiner et al. 1997; Zeiner and Gehring 1995), and the proteasome (Lüders et al. 2000). The known cellular processes in which Bag proteins are involved were recently reviewed comprehensively and are therefore not discussed in detail here (Takayama and Reed 2001).

Whatever the functions of Bag proteins are in the above cellular processes, it is clear from the work of several laboratories that Bag proteins are not strictly essential for the chaperone activity of Hsp70 proteins in folding of non-native proteins (Bimston et al. 1998; Gässler et al. 2001; Lüders et al. 1998; Nollen et al. 2000; Takayama et al. 1997, 1999; Zeiner et al. 1997). This is in contrast to the essential role for GrpE in the chaperone cycle of DnaK. In vivo and in vitro experiments demonstrated that Bag proteins, in particular Bag-1 M, can act as positive and negative regulatory factor for Hsp70 (Bimston et al. 1998; Gässler et al. 2001; Nollen et al. 2000; Takayama et al. 1997; Zeiner et al. 1997). Since the positive or negative regulatory function of Bag-1 depends on its concentration, which in most cell types is rather low (3% of the Hsc70 concentration; Kanelakis et al. 1999), it might be that the inhibitory effects of Bag-1 are restricted to special environmental and metabolic conditions of the cell. An attractive hypothesis is that in the cell, most Bag proteins are associated with their protein partners and exert their regulatory role only locally. Furthermore, it was shown that Bag-1, by accelerating nucleotide exchange, promotes substrate release thereby counteracting the activity of JDPs (Gässler et al. 2001). In the context of Bag-1 binding to the proteasome, Bag-1 could act as substrate discharging factor stimulating substrate release by Hsp70 at the location were the misfolded protein is to be delivered for degradation. Similarly, Bag-1 bound to receptors or kinases could stimulate the premature dissociation of Hsp70 thereby influencing the chaperone-mediated activation of these signaling molecules.

Additional Hsp70 cofactors: cooperation with other chaperones and the degradation machinery

The 35-kDa protein Chip interacts with Hsc70's substrate binding domain and thereby negatively influences the chaperone activity of Hsc70 (Ballinger et al. 1999). This cofactor recently turned out to be a ubiquitin E3 isopeptide ligase promoting the degradation of proteins that were also shown to be substrates of Hsc70 (Hatakeyama et al. 2001; Jiang et al. 2001). The hypothesis is therefore that Chip's interaction with the Hsp70 chaperone systems influences the decision whether a misfolded or regulatory protein is to be refolded into the active state or degraded by the proteasome (Connell et al. 2001; Meacham et al. 2001). The proteasome-interacting Bag-1 may be part of this decision-making system. In addition, Chip seems to have degradation-independent regulatory functions. The heat shock factor HSF1 was found to be activated by the interaction with Chip and translocated into the nucleus in complex with Chip and Hsp70 to activate transcription of heat shock genes (Dai et al. 2003).

The homodimeric 43-kDa Hsp70 interacting protein Hip was originally found in a two-hybrid screen using the Hsc70 ATPase domain as bait (Höhfeld et al. 1995; Velten et al. 2000). This Hsp70 cofactor was subsequently found to be part of the multi-chaperone folding machine that regulates steroid hormone receptors (Prapapanich et al. 1996). Although Hip was proposed to aid chaperone function of Hsp70s by slowing down their nucleotide exchange (Höhfeld and Jentsch 1997; Höhfeld et al. 1995; Lüders et al. 1998) more recent investigations could not provide evidence for such a function of Hip and its biochemical function is still unclear (M.P. Mayer, unpublished results). Rather then having a direct effect on the ATPase cycle of Hsc70 Hip seems to compete with Bag proteins for binding to the ATPase domain of Hsc70 and thereby counteracts the effect of Bag (Kanelakis et al. 2000) (see Fig. 3).

The homodimeric 60-kDa Hsp70–Hsp90 organizing protein Hop (yeast Sti1) was also found in the steroid hormone receptor chaperone complexes (Smith et al. 1993). Hop was shown to promote the assembly of the Hsp70–Hsp90–substrate complex with steroid hormone receptors (Fig. 3).

Chip, Hip, and Hop contain several tetratricopeptide repeat (TPR) domains that constitute protein–protein interaction motifs. The TPR motifs of Hop and Chip bind the EEVD motifs that are found at the C terminus of all known eukaryotic cytosolic Hsp70s and Hsp90s (Scheufler et al. 2000).

Viral control of production and localization of Hsp70s

Increase in Hsp70 chaperone levels following viral infection of cells has been widely observed (see Table 1) (Jindal and Malkovsky 1994). These observations prompted the questions how viruses influence the production of Hsp70 proteins, how specific this induction is and whether such an increase in Hsp70 chaperone levels is advantageous for viral proliferation. From the known regulatory circuits three alternative mechanisms can be inferred. First, the induction may occur indirectly through the production of a large number of proteins that are in an unfolded, aggregation-prone state. In the cytosol or nucleus these protein species could compete with the HSF for binding to Hsp70 thereby leading to an increase of free and active HSF that elicits the transcription of many heat shock genes in the course of the normal cellular stress response (Morimoto et al. 1994). In the ER misfolded

Table 1 Functional interactions of viruses with the Hsp70 system

Virus group	Infecting virus, viral protein	Component of the Hsp70 system	Functional interaction	References
dsDNA viruses				
Polyomaviridae	SV40, TAg	Hsp70	Induction of *hsp*70 transcription	Damania et al. 1998a, 1998b; Khandjian and Turler 1983; Kingston et al. 1986; Sainis et al. 1994; Simon et al. 1988
	SV40, TAg	Hsp70/ Hsc70	Translocation of T-Ag into nucleus	Yang and DeFranco 1994
	SV40, TAg	Hsp70/ Hsc70	Genome replication	Campbell et al. 1997; Li et al. 2001; Sullivan et al. 2000b; Weisshart et al. 1996
	SV40, VP1,2,3	Hsc70	Virion assembly	Chromy et al. 2003; Cripe et al. 1995
	SV40, TAg	Hsc70	Transformation	DeCaprio 1999; Kroll 2002; Sullivan et al. 2000a, 2000b
	JCV	Bag1, Hsc70	Viral transcription	Devireddy and Pater 2000
Papilomaviridae	HPV-11, E1, E2	Hsp70, Hdj1, Hdj2	Genome replication	Lin et al. 2002; Liu et al. 1998
Adenoviridae	AdV, E1A, E1B	Hsp70	Induction of *hsp70/hdj*1 transcription, mRNA export	Agoff and Wu 1994; Glotzer et al. 2000; Herrmann et al. 1987; Kao et al. 1985; Kao and Nevins 1983; Kraus et al. 1994; Lum et al. 1992; Milarski and Morimoto 1986; Moore et al. 1987; Nevins 1982; Phillips and Morimoto 1991; Simon et al. 1987; Theodorakis and Morimoto 1987; Williams et al. 1989; Wu et al. 1986; Yang et al. 1996
	AdV hexon	Hsp70, Hsc70	Virion disassembly?	Niewiarowska et al. 1992; Saphire et al. 2000
	AdV penton	Bag3	Virion disassembly?	Chroboczek et al. 2003
	AdV E1A	Hsc70	?	White et al. 1988
	AdV fiber protein	Hsp70	Virion assembly?	Macejak and Luftig 1991

Table 1 (continued)

Virus group	Infecting virus, viral protein	Component of the Hsp70 system	Functional interaction	References
Herpesviridae	HSV-1; VZV; HCMV, IE1, IE2; EBV	Hsp70, BiP	Induction of *hsp70/grp78* transcription	Caswell et al. 1993; Cheung and Dosch 1993; Colberg-Poley and Santomenna 1988; Colberg-Poley et al. 1992; Hagemeier et al. 1992; Kobayashi et al. 1994; LaThangue et al. 1984; Mao et al. 2001; Phillips and Morimoto 1991; Santomenna and Colberg-Poley 1990
	HSV-2; VZV; HCMV	Hsp70	Nuclear localization	LaThangue and Latchman 1987; Ohgitani et al. 1998, 1999
	HSV-1, UL9	Hsp70, Hsp40, hTid1	Genome replication	Eom and Lehman 2002; Tanguy Le Gac and Boehmer 2002
	HCMV	Hsp70/Hsc70, Bag1	Viral transcription	Niyaz et al. 2003; Takahashi et al. 2001
	EBV, EBNA-LP	Hsp70, Hsc70	Transformation	Kitay and Rowe 1996
Poxviridae	Vaccinia virus	Hsp70	Induction of hsp70 transcription	Brum et al. 2003; Jindal and Young 1992; Sedger et al. 1996; Sedger and Ruby 1994
Siphoviridae	λ-phage	DnaK	Induction of dnaK transcription	Drahos and Hendrix 1982; Tilly et al. 1989
	λP	DnaK, DnaJ, GrpE	DNA replication	Alfano and McMacken 1989a, 1989b; Dodson et al. 1986, 1989; Hoffmann et al. 1992; LeBowitz and McMacken 1984; Liberek et al. 1988, 1990; Wyman et al. 1993; Yochem et al. 1978; Zylicz et al. 1983, 1988; 1989
Myoviridae	P1-phage	DnaK, DnaJ, GrpE	DNA replication	Dibbens et al. 1997; Tilly and Yarmolinsky 1989; Wickner 1990, Wickner et al. 1991a, 1991b, 1992
ssDNA viruses				
Parvoviridae	Parvovirus, H1	Hsp70	Induction of Hsp70 release	Moehler et al. 2003
dsRNA viruses				
Reoviridae	Rotavirus, VP4, VP7	Hsc70	Cell entry	Guerrero et al. 2002; Zarate et al. 2003
	Rotavirus, NSP4		Induction of grp78 transcription	Xu et al. 1998
	Rotavirus, VP7	BiP	Capsid protein folding	Mirazimi and Svensson 2000
	Reovirus	Hsp70	Capsid protein folding	Leone et al. 1996

Table 1 (continued)

Virus group	Infecting virus, viral protein	Component of the Hsp70 system	Functional interaction	References
(+)-strand ssRNA viruses				
Picornaviridae	HCV, E2	BiP	Induction of the UPR	Liberman et al. 1999
	HCV, E1, E2	BiP	Envelope protein maturation	Choukhi et al. 1998
	CAV-9	BiP	Cell entry	Triantafilou et al. 2002
	Poliovirus, CBV-1, P1	Hsp70	Virion assembly ?	Macejak and Sarnow 1992
Togaviridae	Sindbis virus E1, E2	BiP	Envelope protein maturation	Carleton and Brown 1996, 1997; Mulvey and Brown 1995
Closteroviridae	BYV	vHsp70	Cell-to-cell movement	Agranovsky et al. 1998; Alzhanova et al. 2000, 2001; Peremyslov and Dolja 2002; Peremyslov et al. 1999
	BYV, CTV	vHsp70	Virion assembly	Napuli et al. 2000; Satyanarayana et al. 2000, 2004
Tobamoviridae	TMV, CP	Hsp70	Induction of heat shock response	Jockusch et al. 2001
(−)-strand ssRNA viruses				
Bunyaviridae	Topsovirus NSm	JDP	Cell-to-cell movement	Soellick et al. 2000
Rhabdoviridae	Rabies virus; VSV	Hsc70	Present in virion particle	Sagara and Kawai 1992
	VSV-G	BiP	Envelope protein folding/maturation	de Silva et al. 1990, 1993; Hammond and Helenius 1994; Machamer et al. 1990
Paramyxoviridae	NDV; SV5, HN; measles virus; sendai virus	BiP	Induction of UPR	Bolt 2001; Collins and Hightower 1982; Ng et al. 1989; Peluso et al. 1978; Watowich et al. 1991
	SV5, HN, SV5, F	BiP	Envelope protein folding/maturation	Ng et al. 1989; Parks and Lamb 1990; Roux 1990; Tomita et al. 1999; Watowich et al. 1991
	CDV	Hsp70, Hsc70	Capsid protein folding	Oglesbee and Krakowka 1993; Oglesbee et al. 1990; Zhang et al. 2002
	CDV, measles virus	Hsp70, Hsc70	Genome replication	Oglesbee et al. 1996; Zhang et al. 2002
Orthomyxoviridae	Influenza A, H	BiP	Envelope protein folding/maturation	Hogue and Nayak 1992; Singh et al. 1990
	Influenza A	Hsc70	Present in virion particle	Sagara and Kawai 1992
	Influenza A	$p58^{IPK}$, Hsc70	General regulation of translation	Melville et al. 1997, 1999; Yan et al. 2002

Table 1 (continued)

Virus group	Infecting virus, viral protein	Component of the Hsp70 system	Functional interaction	References
Retroidviruses				
Retroviridae	RSV	BiP	Induction grp78 transcription	Peluso et al. 1978; Stoeckle et al. 1988
	HTLV-1	Hsc70	Cell-to-cell movment	Fang et al. 1999; Sagara et al. 1998
	HTLV-1, Tax	Hsp70, hTid1	Transformation	Cheng et al. 2001
	HIV, gp120	Hsp70	Induction of stress response	Furlini et al. 1994; Wainberg et al. 1997
	HIV	Hsp70	Nuclear import of pre-integration complex	Agostini et al. 2000
	HIV, Tat	Hsp70	Viral transcription	O'Keeffe et al. 2000
	HIV, gp160	BiP	Envelope protein maturation	Earl et al. 1991; Otteken et al. 1996
	HIV	Hsp70, Hsc70	Present in virion particle	Gurer et al. 2002
Hepadnaviridae	HBV, Pol	Hsp70, Hsp40	Reverse transcription	Beck and Nassal 2001, 2003; Gyoo Park et al. 2002; Hu and Anselmo 2000; Hu and Seeger 1996; Hu et al. 2002; Hu et al. 1997
	HBV, L	Hsp70	Envelope protein maturation	Cho et al. 2003; Lambert and Prange 2003; Löffler-Mary et al. 1997; Prange et al. 1999

Abbreviations. Viruses: AdV, adenovirus; BYV, beet yellow virus; CAV-9, coxsackievirus A9; CDV, canine distemper virus; CTV, citrus tristeza virus; EBV, Epstein–Barr virus; HBV, hepatitis B virus; HCMV, human cytomegalovirus; HCV, hepatitis C virus; HIV, human immunodeficiency virus; HPV, human papilomavirus; HSV, herpes simplex virus; HTLV-1, human T cell leukemia virus 1; JCV, JC virus; NDV, Newcastle disease virus; RSV, Rous sarcoma virus; SV5, simian virus 5; SV40, simian virus 40; TMV, tobacco mosaic virus; VSV, vesicular stomatitis virus; VZV, varicella zoster virus. Proteins: HN, haemagglutinin-neuraminidase; TAg, large T antigen; vHsp70, virus encoded Hsp70; BiP/Grp78, immunglobulin binding protein/glucose regulated protein 78 kDa (ER Hsp70); all other abbreviation name viral proteins

proteins would bind to the ER Hsp70 homolog BiP/Grp78 and thereby elicit the unfolded protein response that increases the transcription of genes encoding for ER resident chaperones (Urano et al. 2000). Second, the viruses could interfere with stress signal transduction pathways upstream or independent of HSF thereby using pathways that are independent of unfolded proteins inside the cell. Third, specialized viral proteins may induce directly the increased production of Hsp70 proteins, either by binding to the promoters of specific Hsp70-encoding genes or by stabilizing certain mRNAs and activating their translation. As detailed in the following parts, all three mechanisms have been observed. While the changes in the levels of Hsp70 proteins were investigated in many systems, changes in the levels of co-chaperones of Hsp70 may be inferred but are generally not mentioned.

Induction of the stress response

The stress response to unfolded proteins is apparently the reason for the induction of heat shock gene transcription during infection with the varicella zoster virus (VZV, *Herpesviridae*). Inclusion bodies, which are formed in the nucleus of the host cells in the course of the virus infection, sequester Hsp70 and presumably liberate HSF for heat shock gene transcription (Ohgitani et al. 1998). More direct evidence for this mode of viral heat shock response induction was provided by Rajes and coworkers (Jockusch et al. 2001). Analyzing Hsp70 and Hsp18 mRNA and protein levels in tobacco leaves after infection with wild-type tobacco mosaic virus (TMV, *Tobamoviridae*) or a mutant virus, which encodes a temperature sensitive coat protein, they found that degree and time of induction paralleled the amount of insoluble viral coat protein suggesting that the amount of unfolded protein is the inducing agent.

Such an induction mechanism was also found for the ER resident Hsp70 and Hsp90 homologs, BiP/Grp78 and Grp94, which are overproduced in the course of a hepatitis C virus (HCV, *Picornaviridae*) infection. The HCV envelope protein E2 is retained in a pre-Golgi compartment in an at least partially misfolded state and bound by BiP (Liberman et al. 1999). Since BiP is part of the sensing system that detects misfolding in the ER, the interaction of E2 with BiP activates the feedback mechanism of the unfolded protein response leading to the overproduction of a number of ER resident chaperones including BiP and Grp94 (Kozutsumi et al. 1988). Similarly, the BiP encoding gene is induced during the infection with the paramyxoviruses measles virus, Sendai virus and simian virus 5 (SV5) (Bolt 2001; Peluso et al. 1978). However, the inducing viral component, the hemagglutinin–neuraminidase glycoprotein, is not retained in the ER in an unfolded state. The high flux of this protein through the ER seems to suffice for the induction of the unfolded protein response (Ng et al. 1989; Watowich et al. 1991). The case of the rotavirus (*Reoviridae*) nonstructural glycoprotein NSP4, which also induces the production of BiP and Grp94, seems to be slightly different, since an interaction with BiP or Grp94 could not be detected. The mechanism of this induction is not clear (Xu et al. 1998).

Interference with stress signaling pathways is described for the human immune deficiency virus (HIV). HIV induces the synthesis of *hsp70* and *hsp27* mRNA through the interaction of its envelope glycoprotein gp120 with cell surface receptors (Furlini et al. 1994; Wainberg et al. 1997). Isolated gp120 is also able to induce the heat shock genes in the absence of intact virus particles. The Epstein–Barr virus also seems to induce the *hsp70* and *hsp90* genes through the interference with signal transduction pathways since the induction is clearly dependent on virus attachment but independent of viral protein synthesis and involves an influx of Ca^{2+} (Cheung and Dosch 1993).

Selective *hsp* gene induction

A specific *hsp* induction is realized by a number of viruses, e.g., adenovirus (AdV), herpes simplex virus (HSV), and simian virus 40 (SV40) and is characterized by the selective induction of specific chaperone-encoding genes while the expression of other heat shock genes is not influenced. The transactivating viral proteins thereby recognize the specific promoters directly by binding to *cis*-acting DNA elements or by interacting with promoter-specific transcription factors. The SV40 large T-antigen (TAg), for example, induces the *hsp70* gene expression (Kingston et al. 1986; Sainis et al. 1994; Simon et al. 1988) by binding to the general transcription initiation factors, the TATA-binding protein (TBP) and TFIIA, and stabilizing the transcription pre-initiation complex at the *hsp70* promoter TATA-element, but not at two other promoters tested (Damania et al. 1998a; Gilinger and Alwine 1993). The adenoviral early gene products E1A and E1B synergistically stimulate *hsp70* gene expression (Herrmann et al. 1987; Kao et al. 1985; Phillips et al. 1991; Wu et al. 1986) in a cell cycle specific manner (Milarski and Morimoto 1986; Simon et al. 1987). This stimulation is mediated through an interaction of E1A with the CCAAT-box binding factor (Agoff and Wu 1994; Lum et al. 1992; Williams et al. 1989) and the disruption of the inhibitory complex of DR1 with TBP (Kraus et al. 1994). Furthermore, AdV specifically promotes the nuclear export of the *hsp70* mRNA by interaction of the mRNA with E1B, thereby circumventing the general virus-induced export block for cellular mRNAs (Moore et al. 1987; Yang et al. 1996). Late during AdV infections the *hsp70* mRNA levels decline precipitously while transcription continues demonstrating the complexity of the viral regulation of *hsp70* expression (Theodorakis and Morimoto 1987).

In addition to the induction of production of Hsp70, viruses are able to influence the localization of the Hsp70 systems. The avian AdV CELO gene product Gam1 for example leads to a strong accumulation of Hsp70 and its co-chaperone Hdj1 in the nucleus (Glotzer et al. 2000). Similarly, the human cytomegalovirus (HCMV) directs Hsp70 to the nucleus in the early phase of infection but redirects it to the cytoplasm in the late phase (Ohgitani et al. 1999).

What is the significance of the virus induced *hsp70* expression? In view of the fact that transcription of most cellular genes is shut down and mRNA processing, transport and translation is regulated by many viruses, the virus-induced increase of Hsp70 levels in the cell seems to be important for the viral life cycle. Two clear examples where the essentiality of the Hsp70 system for virus proliferation was demonstrated beyond doubt are the bacteriophage λ and the avian AdV CELO. The *E. coli* Hsp70 homolog DnaK (GroPC) was originally identified as host factor essential for bacteriophage λ proliferation and loss of function mutations in *dnaK*, *dnaJ*, and *grpE* can only be complemented by transduction of the respective gene (Georgopoulos 1977; Saito and Uchida 1977). The CELO inducer of *hsp70* and *hdj1* transcription, Gam1 is also essential for virus proliferation. However, Gam1 can be replaced by a heat shock, being most efficient immediately before infection, or the transduction of the *hdj1* gene (Glotzer et al. 2000).

Functions of the Hsp70 system in the viral life cycle

Hsp70 systems are potentially involved in all phases of the viral life cycle including cell entry, virion disassembly, the transfer of viral genome into the nucleus, replication of the viral genome, morphogenesis of the virion particles, and transformation of the cell. The following discussion is ordered according to these stages of the life cycle rather than to virus systematics and the same viruses are discussed in several sections in different contexts. For a systematic summary see Table 1.

Cell entry: interaction of viruses with cell-surface exposed Hsp70s

Before animal viruses can transfer their virion particles into host cells by fusion of their envelope with the plasma membrane or by receptor-mediated endocytosis, they must attach to specific surface receptors. This surface attachment is, at least in some cases, a multistep process that can involve surface-exposed Hsp70 proteins. Although Hsp70 and Hsc70 do not contain export signal sequences and more importantly depend in their chaperone function on repetitive cycles of ATP hydrolysis, they are found on the cell surface of a number of different cell types including tumor cells (Hantschel et al. 2000; Kaur et al. 1998; Multhoff et al. 1998; Multhoff and Hightower 1996; Shin et al. 2003), virus infected cells (Brenner and Wainberg 1999; Chouchane et al. 1994), spermatogenic cells (Boulanger et al. 1995; Miller et al. 1992), epidermal cells (Rocchi et al. 1993), arterial smooth muscle cells (Johnson and Tytell 1993), monocytes and B cells (Manara et al. 1993; VanBuskirk et al. 1989, 1991). It is not clear in each case how Hsc70 and Hsp70 locate to the cell surface, however, for antigen-presenting cells it was shown that specific receptors exist that bind Hsp70 and Hsc70 for uptake out of the extracellular space (Arnold-Schild et al. 1999; Asea et al. 2002; Becker et al. 2002; Binder et al. 2000; Lipsker et al. 2002; Sondermann et al. 2000). This uptake is proposed to signal necrotic and lytic processes and to induce an inflammatory response (Dybdahl et al. 2002; Milani et al. 2002; Vabulas et al. 2002). Furthermore, it was proposed that Hsp70-bound peptides are transferred in the endosomal pathway onto MHC II molecules for antigen presentation (Becker et al. 2002; Domanico et al. 1993; Kishi et al. 2001; Nicchitta 2000; Panjwani et al. 1999; Pierce 1994; Roth et al. 2002; Suzue et al. 1997; VanBuskirk et al. 1989).

The nonenveloped rotavirus, a double-stranded RNA virus, needs cell surface-exposed Hsc70 at a postattachment step for the successful infection of epithelial cells (Arias et al. 2002). Treatment of the epithelial cells with monoclonal antibodies against Hsc70 prevented internalization and infection by rotavirus but did not obliterate its attachment to the cells, while infection with poliovirus and reovirus was not affected by the antibody treatment (Guerrero et al. 2002). Hsc70 was shown to bind to free rotavirus particles and preincubation of the virus particles with Hsc70 significantly decreased their infectivity (Guerrero et al. 2002). The C-terminal half of the viral coat protein VP5 was demonstrated to be sufficient for interaction with Hsc70 and the interaction of VP5 as well as whole virus particles with Hsc70 could be competed with a 17-mer peptide containing a sequence stretch of the VP5 C-terminal domain (Zarate et al. 2003). These observations pose interesting questions regarding the mechanism of this interaction: What is the role of Hsc70 in the internalization of rotavirus? Are substrate-binding by Hsc70 and its chaperone activity involved?

In view of the fact that Hsc70 proteins loaded with substrate peptides are internalized through receptor-mediated endocytosis, it is tempting to speculate that the internalization process may depend on, or even may be triggered by, peptide binding to the substrate-binding domain of Hsc70. Rotavirus could use such a mechanism for efficient cell entry by providing the substrate signal to Hsc70. Since there is no ATP available for cell surface exposed Hsc70, the interaction of the substrate-binding domain of Hsc70 with a substrate polypeptide would be a very slow reaction ($\tau \approx 100$–1,000 s; Pierpaoli et al. 1997; Takeda and McKay 1996). These limitations, however, do not preclude such a mechanism for the interaction since attachment steps prior to the interaction with Hsc70 are necessary for rotavirus internalization providing the time window and the high local concentration necessary for an efficient binding. Since substrate dissociation rates are very low in the ADP or nucleotide-free state, the rotavirus would be firmly attached to the cell once associated to Hsc70. Such a mechanism could be tested by competing rotavirus infection with high concentrations of peptides, which are known to have a high affinity for Hsc70 and are unrelated to the VP5 sequence, or by replacing surface exposed wild-type Hsc70 by mutant proteins, which are defective in their affinity for substrates (compare Mayer et al. 2000b). Alternatively, rotavirus could bind to any other part of the Hsc70 molecule without involving the chaperone mechanism of Hsc70. For example a co-chaperone binding site could serve as interaction site, as is the case for the Hsc70 receptor CD40, which competes with the co-chaperone Hip for binding to the ATPase domain of Hsc70 (Becker et al. 2002). In favor of this mode of binding is the fact that the peptide, which was able to compete with rotavirus infection, does not seem to be a high-affinity substrate for Hsc70 as judged from the sequence of the peptide using the DnaK algorithm and the BiP scoring system (Blond-Elguindi et al. 1993; Rüdiger et al. 1997). Rotavirus would then be internalized as a hitch-hiker on Hsc70. A more extensive function of the cell-surface exposed Hsc70, for example by chaperoning a viral coat protein to promote translocation through the membrane or uncoating of the virion, seems unlikely because Hsc70 depends in its chaperone function on ATP and the assistance of a JDP both of which have not yet been found on the cell surface. At later stages, however, in endosomal vesicles, such a chaperone action may be possible, because ATP as well as co-chaperones are present in the ER and could be transferred by vesicular flow to endocytotic vesicles.

Similar to rotavirus, the coxsackievirus A9 (CAV-9), a nonenveloped RNA virus of the *Picornaviridae* family, also interacts with a cell-surface exposed Hsp70 homolog. Its target Hsp70, the primarily ER resident BiP/Grp78, was shown by fluorescence energy transfer experiments to be associated to major histocompatibility complex (MHC) I molecules (Triantafilou et al. 2002). BiP is known to interact with MHC molecules already in the ER supporting their folding and assembly (Paulsson and Wang 2003). It therefore may occasionally remain in complex with the MHC proteins and migrate with them to the plasma membrane. Monoclonal BiP-specific antibodies prevent attachment and cell entry of CAV-9 (Triantafilou et al. 2002).

Another example of a virus interacting with Hsp70 proteins on the cell surface is the retrovirus human T lymphotropic virus type 1 (HTLV-1). The cell-free infectivity of this virus is very low (Clapham et al. 1983) but as in many other retrovirus infections, close cell-to-cell interactions between HTLV-1 harboring cells and target cells leads to syncytium formation allowing direct cell-to-cell transfer of the virus (Hoshino et al. 1983). Cell-surface exposed Hsc70 is essential for this process as shown by blocking syncytium formation with Hsc70-specific monoclonal antibodies or a peptide derived from the sequence of the HTLV-1 glycoprotein gp46, which binds to Hsc70 (Fang et al. 1999; Sagara et al. 1998). This peptide seems to be a good Hsc70 substrate as judged by the DnaK-algorithm (Rüdiger et al. 1997) and syncytium formation could therefore involve the substrate binding properties of Hsc70.

Uncoating and genome release

The viral genomes are packaged in nucleocapsid structures which allow condensation of the viral nucleic acid in a very small space and serve as protective coat against a hostile environment. For successful infection the viral genome has to be released from the virion particle either by disassembly of the coat or by opening of a pore. The coats, however, must be stable to prevent genome release outside of a host cell and uncoating of the virion particle is most likely a thermodynamically unfavorable process. The differences between extra- and intracellular environment, oxidizing versus reducing conditions and the different ionic milieu, which are experienced by the capsid of nonenveloped viruses, may not be sufficient for the destabilization of the virion particle. If the virion particle would be stable in the extracellular but unstable in the intracellular milieu, thermodynamics would require that the assembly reaction in the cytoplasm should be unfavorable. Therefore, viruses for which the uncoating does not occur in endocytic vesicles aided by acidification may involve cellular components in the genome release process. Since Hsp70 chaperones are known to be involved in the disassembly of oligomeric protein structures, the best example of which is the uncoating of clathrin coated vesicles (Greene and Eisenberg 1990; Ungewickell 1985), it is conceivable that the Hsp70 chaperone machinery assists the uncoating of virion particles.

Such a process was described for an AdV. Soon after the release of the virion particle from endocytic vesicles into the cytoplasm, Hsp70 and Hsc70 can be found attached to the hexon protein, the major AdV coat proteins (Niewiarowska et al. 1992). In addition, Hsp70 and its co-chaperone Bag3 interact with the penton protein, the base and fiber-forming virion component (Chroboczek et al. 2003). The intact nucleocapsid is transported to the nuclear pore complex using the normal nuclear localization signal (NLS)-dependent nuclear import machinery as demonstrated by competition with classical NLS-containing proteins and inhibition with nuclear import inhibitors like GTPγS (Saphire et al. 2000). The nucleocapsid docks with the nuclear pore by interaction of its hexon protein with components of the pore complex without transiting into the nucleus. The viral DNA is subsequently transferred into the nucleus in an Hsp70-dependent manner. Since purified hexon can enter the nucleus in an Hsp70-independent manner while the viral DNA cannot (Saphire et al. 2000), it is plausible that the nucleocapsid, which is too large to pass through the nuclear pore complex, is disassembled in an Hsp70-dependent manner allowing the viral DNA to enter the nucleus. The final proof for such a mechanism is, however, still lacking because in a reconstituted import assay containing the necessary import factors for hexon import supplemented with Hsp70 the viral DNA was not transferred into the nucleus to any significant extent. The inability of the reconstituted system used to uncoat the virion particle was most likely due to the absence of Hsp70 co-chaperones such as JDPs or Bag-domain proteins. Similar to the uncoating of clathrin coated vesicles where the clathrin associated JDP auxilin targets Hsc70 to clathrin for the multiple ATPase cycles requiring uncoating reaction, JDPs may be necessary for targeting Hsp70 proteins to the AdV coat and disassembly may require multiple J-domain stimulated rounds of ATP hydrolysis. Furthermore, proteolytic processes including the proteasome may also be involved since three ubiquitin-protein isopeptide ligases were found associated with the virion (Chroboczek et al. 2003).

Replication and reverse transcription

The involvement of Hsp70 systems in viral DNA replication was first demonstrated by genetic and biochemical means for the *E. coli* bacteriophage λ (Alfano and McMacken 1989b; Georgopoulos 1977; Georgopoulos and Herskowitz 1971; Mensa-Wilmot et al. 1989; Saito and Uchida 1977; Zylicz et al. 1989; for review see Zylicz et al. 1999). The bacteriophage protein λP sequesters the *E. coli* DNA helicase DnaB and recruits it to the origin of replication oriλ of the bacteriophage genome, where it interacts with the four dimers of λO assembled at oriλ to form a multimeric complex (Dodson et al. 1985; Liberek et al. 1988; Roberts and McMacken 1983). In the absence of the Hsp70 system of *E. coli*, DnaK, DnaJ and GrpE, the λO·λP·DnaB complex is stalled at the oriλ and DNA unwinding and therefore replication cannot start. The most likely reason for this block is that the affinity of λP to DnaB is significantly higher (at least fivefold) than the affinity between the *E. coli* replication initiation factor DnaC and DnaB (Mallory et al. 1990). λP therefore outcompetes DnaC for binding to DnaB with the consequence that the λP·DnaB complex cannot spontaneously disassemble. The thermodynamically stable complex of λO, λP, and DnaB at oriλ is disassembled by the chaperone action of DnaJ and DnaK, which interact with and sequester λP in an ATP-dependent process thereby liberating DnaB for the unwinding of the DNA and the initiation of replication (Alfano and McMacken 1989a; Dodson et al. 1989; Hoffmann et al. 1992; Liberek et al. 1988; Zylicz et al. 1989). If DnaK is in large excess over λP the nucleotide exchange factor GrpE is not essential. However, if DnaK is at more stoichiometric concentrations GrpE enhances the efficiency of the process greatly, demonstrating that nucleotide exchange and therefore multiple rounds of the ATPase cycle with substrate binding and release is central to the efficiency of the disassembly reaction (Alfano and McMacken 1989a; Alfano and McMacken 1989b; Zylicz et al. 1988, 1989).

For the replication of the P1 phage genome, as well as for the replication of the F-plasmid, the DnaK/DnaJ/GrpE-system plays a different role (Kawasaki et al. 1990; Tilly and Yarmolinsky 1989; Wickner et al. 1991a; , 1991b, 1992; Wickner 1990). The initiator proteins RepA for P1 phage and RepE for the F-plasmid form stable homodimers in vivo and in vitro (Ishiai et al. 1994; Swack et al. 1987; Wickner et al. 1991b). RepA does not bind DNA in its dimeric form, while RepE binds to an inverted repeat sequence motif, the operator of the *repE* gene repressing its transcription (Ishiai et al. 1994). Both proteins can only bind in their monomeric form to the direct repeat sequence motifs, the so-called iterons, within their respective origin of replication to initiate DNA replication. The conversion into monomers requires the chaperone action of DnaK, DnaJ, and GrpE, as demonstrated in vitro and by mutant analysis in vivo (Dibbens et al. 1997; Ishiai et al. 1994; Wickner et al. 1991b, 1992). A constitutively monomeric mutant protein of RepE is able to initiate replication of the P1 genome independent of the DnaK system (Matsunaga et al. 1997). The crystal structure of the constitutively monomeric mutant protein of RepE in complex with its iteron DNA suggested that large conformational changes are necessary for the dimer–monomer interconvertion explaining the necessity of Hsp70 action (Komori et al. 1999; Sharma et al. 2004).

The involvement of the Hsp70 system in genome replication has also been demonstrated for eukaryotic viruses. The binding of the human papillomavirus-11 (HPV) DNA helicase E1 to DNA is enhanced by Hsp70, Hdj1, and Hdj2, whereby the action of Hsp70 is ATP dependent (Liu et al. 1998). Hdj2 stabilizes E1 in a dihexameric state on supercoiled and relaxed DNA. In the presence of topoisomerase I and single-stranded DNA-binding protein E1 unwinds supercoiled DNA. This unwinding reaction, however, was independent of an origin of replication. The HPV origin binding protein E2, which is essential for the formation of the pre-initiation complex at the origin of replication but dispensable for the elongation reaction, binds with high affinity to E1 and inhibits the unwinding reaction. This inhibition is abrogated by the Hsp70 chaperone system (Lin et al. 2002). Together these data suggest the following model. The E2 protein recognizes the origin of replication of the HPV genome and tethers the E1 helicase to the origin DNA. The Hsp70 chaperone machinery enhances this assembly process significantly, maybe by remodeling the hexameric, ring-shaped E1 to allow efficient DNA threading. The Hdj-proteins seem to stabilize a dihexameric form of E1 probably in preparation for bidirectional unwinding. Finally, Hsp70 and Hdj-proteins remove the E2 protein in a reaction similar to the *E. coli* DnaK machinery removing the λP protein from the pre-initiation complex. Consequently, the block on the stalled complex is lifted allowing the start of DNA unwinding by E1 and the action of DNA primase (Lin et al. 2002).

The Hsp70 chaperone machinery also seems to be essential for replication of the SV40 genome. The SV40 TAg is a multifunctional protein that among other functions acts as replication initiator protein and DNA helicase. It contains at its N terminus the signature motif for JDPs, the J-domain (Fig. 4). Many point mutations or deletions within this domain lead to a reduction or even loss of helicase function (Campbell et al. 1997; Li et al. 2001; Weisshart et al. 1996). SV40 proteins with a deletion of the N-terminal J-domain does not assemble into the correct hexameric structure and exhibits a lower affinity for the SV40 origin of replication (Weisshart et al. 1996). The J-domain also seems to contribute to the interaction with DNA polymerase α-primase (Dornreiter et al. 1990).

The DNA-replication of HSV is also affected by Hsp70 chaperones. The affinity of the dimeric replication initiator protein UL9 to oriS, one of the viral origins of replication, is greatly stimulated by Hsp70 and Hdj1 (Tanguy Le Gac and Boehmer 2002). This effect is largely due to an increase in association rate whereas the dissociation rate is seemingly unaltered. In contrast, the Hsp70 chaperone team does not affect the affinity of UL9 to single-stranded DNA as well as the helicase activity. Interestingly, the monomeric C-terminal origin-binding domain of UL9 binds independent of Hsp70 or Hdj1 to oriS with even higher affinity (Tanguy Le Gac and Boehmer 2002). One possible interpretation of these results is that Hsp70 monomerizes UL9 similar to the effect of the *E. coli* DnaK/DnaJ/GrpE system on the replication initiator proteins of P1 phage, RepA, and of the F-plasmid, RepE, thereby enhancing the affinity and specificity of UL9 for oriS. Alternatively, Hsp70 could assist a conformational transition of UL9 that would increase the accessibility of the origin-binding site of UL9 in its C-terminal domain and thereby increase the affinity of dimeric UL9 for oriS. Since the strand-opening reaction is strictly dependent on the amount of UL9 bound to its high and low-affinity sites flanking the central A/T-rich region of oriS, Hsp70 increases with the UL9 occupancy on oriS the efficiency of DNA replication in HSV.

In canine distemper virus (CDV, *Paramyxoviridae*) the replication of the negative strand RNA occurs in a ribonucleocapsid particle containing the virus-encoded proteins N and P, the major core protein and the RNA-dependent RNA polymerase. The observation that the induction of the stress response promoted cytopathic effects of CDV infection and an association of Hsp70 proteins with the nucleocapsid particles suggested a possible contribution of Hsp70 proteins to viral replication (Oglesbee and Krakowka 1993; Oglesbee et al. 1990). Isolation of nucleocapsid particles from stressed and unstressed cells under conditions of ATP depletion demonstrated that Hsp70 and Hsc70 are associated under normal as well as under stress conditions. Furthermore, antibodies against Hsp70 reduced the RNA polymerase activity associated with the nucleocapsid particles, while addition of purified Hsp70/Hsc70 proteins increased the polymerase activity (Oglesbee et al. 1996). Isolation of nucleocapsid particles in the presence of ATP, which led to a depletion of Hsp70 proteins, yielded particle devoid of polymerase activity. These data clearly demonstrate the importance of Hsp70 association with the nucleocapsid particles for viral genome replication and transcription. The mechanism of this interaction is still unknown. Two possible modes of action could be that the Hsp70 proteins interact and remodel the N protein to make the RNA accessible to the polymerase, or they could directly chaperone the polymerase and thereby enhance its activity.

It has also been reported that the reverse transcriptase of the retroid hepadnavirus depends on the Hsp70 folding machinery for activity (Beck and Nassal 2001; Hu et al. 1997). In this case cellular folding processes are used that are normally involved in the control of stability and activity of regulatory proteins such as receptors and protein kinases involved in signal transduction, cell cycle regulation, and apoptosis. In addition to JDP and Hsp70 proteins, these processes involve the Hsp90 chaperones and an increasing number of co-chaperones including Hop, Hip, p23, p50^{cdc37}, and immunophilines (Pearl and Prodromou 2002; Pratt and Toft 2003; Richter and Buchner 2001; Young et al. 2001). The current model of the action of this chaperone machine is derived from the assembly reaction of steroid hormone receptors. The reaction cycle starts with the interaction of Hdj1 and Hsp70 with the so-called client protein cotranslationally or directly after de novo synthesis. The TPR-containing protein Hop assembles Hsp70, Hsp90 and the client in an early complex. p23 and immunophilins replace Hop and Hsp70 to yield the mature complex which dissociates with a half-life of approximately 5 min (Smith 2000). Hdj1 and Hsp70 can rebind the released client to restart the cycle. It is believed that the chaperone activity of Hsp70 brings the client into a certain conformation in which it is captured by the clamp mechanism of Hsp90. In this conformation the client is inactive but rapidly activatable by protein modifications such as phosphorylation or by binding to a ligand or a partner protein. After the initial observation that Hsp90 is associated with the reverse transcriptase of the duck hepatitis B virus (HBV) (Hu and Seeger 1996), a number of in vitro reconstitution experiments demonstrated that the activation of the reverse transcriptase follows the general scheme of the activation of cellular regulatory proteins (Beck and Nassal 2001, 2003; Gyoo Park et al. 2002; Hu and Anselmo 2000; Hu et al. 1997, 2002). Only after interaction with Hdj1, Hsp70, Hop, and Hsp90 in an ATP dependent reaction, is the reverse transcriptase able to assemble with the ϵRNA located near the 5′ end of the pregenome as a template and with the core protein into a nucleocapsid, where DNA synthesis commences with the protein priming reaction and the template switch from the 5′ to the 3′ end (Bartenschlager et al. 1990; Bartenschlager and Schaller 1992; Pollack and Ganem 1994; Tavis et al. 1994; Wang and Seeger 1992, 1993). The co-chaperone p23 increases the efficiency of the process. A more recent study demonstrating that Hdj1 and Hsp70 alone are able to activate the reverse transcriptase supports the notion that Hsp70 transforms the re-

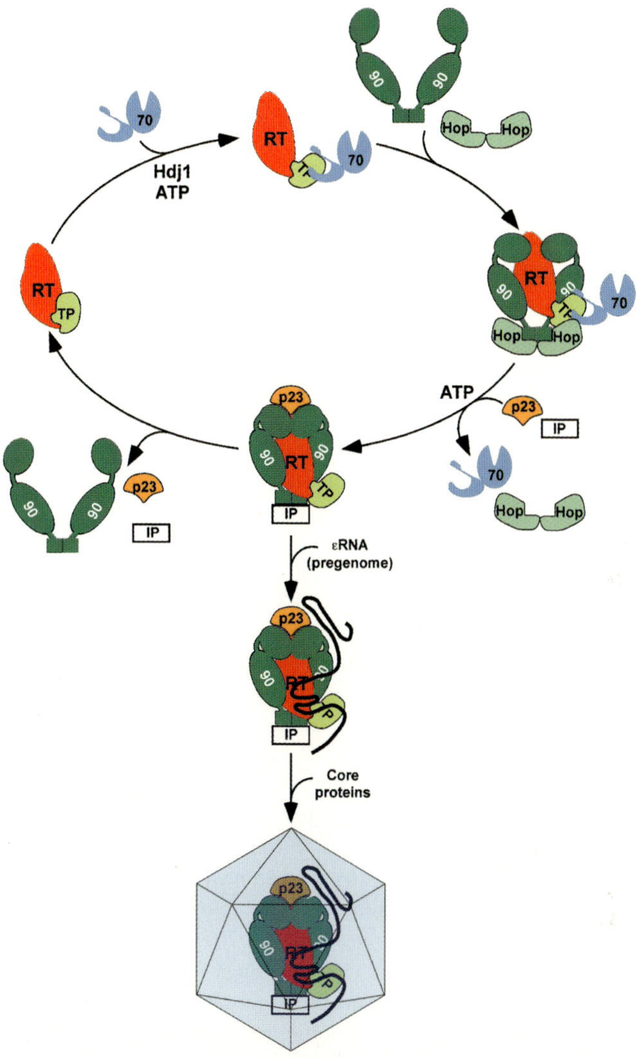

Fig. 5 Activation of the reverse transcriptase/DNA polymerase of hepadnavirus. The activation combines the model for the activation of steroid hormone receptor by Hsp90, the Hsp90 ATPase cycle with proposed conformational changes in Hsp90 and the activation model from Hu et al. (Hu et al. 1997; Mayer et al. 2002; Pearl and Prodromou 2002; Pratt 1997; Richter and Buchner 2001; Smith 2000; Young et al. 2001). *RT*, Hepadnavirus P protein reverse transcriptase/DNA polymerase; *TP*, N-terminal domain of the P protein; *70*, Hsp70; 90, Hsp90 with N-terminal ATPase domain, middle domain and C-terminal domain; *IP*, immunophilins. The interaction sites of the Hsp70 and Hsp90 are not known and therefore arbitrary

verse transcriptase by a modulation of its conformation into the active state in which it is stabilized by Hsp90 and co-chaperones and which allows association with the RNA and the core proteins to form the reverse transcription-competent nucleocapsid particle (Beck and Nassal 2003) (summarized in Fig. 5). Finally, Hsp70 and Hsp90 are found incorporated in the released virus particles supporting the idea that the two chaperones stabilize the reverse transcriptase during transmission in the extracellular space and allow immediate activation after reintroduction into a suitable host cell. Since Hsp70 is also found in other viruses including the retrovirus human immune deficiency virus (HIV) and non-retroviruses of the negative-strand RNA group (rabies virus, vesicular stomatitis virus, influenza A virus, etc.) in similar amounts as the polymerase it is possible that the chaperones assist reverse transcription and RNA-dependent RNA polymerase reactions in other viruses as well (Gurer et al. 2002; Sagara and Kawai 1992).

In summary, Hsp70s are involved in viral genome replication by monomerization of initiator proteins, assisting assembly and disassembly of preinitiation complexes, and by stabilizing and activating helicases and polymerases.

Viral gene expression

Most viruses exploit the cellular transcription and translation machineries for the expression of their genes and therefore recruit initiation and elongation factors. Since some of the involved host factors interact with components of the Hsp70 system, the chaperone system is also important for this stage of the viral life cycle. Several transcription initiation factors interact physically with the Hsp70 co-chaperone Bag1 in vitro and Bag1 stimulates general transcription activity in vitro and when overexpressed in vivo in an Hsp70-dependent manner (Niyaz et al. 2003; Niyaz et al. 2001; Zeiner et al. 1999). Such a stimulation of transcription was also observed when viral promoters were used in reporter gene constructs with promoters of the human polyomavirus JCV and HCMV (Devireddy et al. 2000; Takahashi et al. 2001). The molecular mechanism of the general transcriptional activation, however, is still unclear.

As discussed above, Bag1 stimulates ADP dissociation from Hsp70 proteins. In the presence of physiological ATP concentrations, ATP rebinds rapidly inducing the conformational change in Hsp70 and Hsc70 that leads to substrate release (Gässler et al. 2001; Höhfeld and Jentsch 1997; Sondermann et al. 2001). Taking these properties into account it is possible that Bag1, tethered to DNA by its own unspecific DNA binding activity and/or by interaction with a transcription initiation factor, assists the assembly of transcription initiation complexes by triggering the release of Hsc70-bound transcription factors at the site of the promoter. Alternatively or in addition, Bag1–Hsc70 interaction could remodel initiation complexes thereby stimulating the promoter clearance of the RNA polymerase.

A different way of stimulating virus specific transcription is used by HIV-1. The Tat protein specifically binds to the TAR stem–loop structure at the 5' end of the nascent viral transcript and activates HIV-1 transcription by enhancing the processivity of RNA polymerase (Cullen 1998; Jones 1997). This activation is mediated by the human transcription elongation factor P-TEFb, which directly interacts with Tat and phosphorylates the C-terminal tail of RNA polymerase II (Chen et al. 1999). The protein kinase subunit of P-TEFb is the cyclin-dependent protein kinase Cdk9 which is a Hsp70 and Hsp90 client. Interference with the Hsp70–Hsp90 chaperones, for example using the Hsp90 inhibitor geldanamycin, prevents the formation of the functional Cdk9·cyclin T1 complex and the transcriptional stimulation by Tat (O'Keeffe et al. 2000).

A common problem for gene expression of RNA viruses is that double-stranded RNA intermediates of their replication and gene expression induce the protein kinase PKR (protein kinase-RNA-activated) which shuts down translation by phosphorylating the translation initiation factor eIF2α (Galabru and Hovanessian 1987; Gale et al. 1998). PKR is also induced by interferon and as such it is part of the host defense strategy against viral infections. To circumvent the PKR-mediated block to viral proliferation influenza A virus induces the cellular TPR-domains containing JDP p58^{IPK}, which down-regulate PKR in an Hsp70-dependent manner (Lee et al. 1990, 1992, 1994; Tang et al. 1996; Melville et al. 1997, 1999). In uninfected, unstressed cells p58^{IPK} forms a complex with Hdj1 which is proposed to be the inactive form of p58^{IPK} (Melville et al. 1997, 1999). During influenza A virus infection the amount of Hdj1 that co-precipitates with p58^{IPK} first increases about twofold and then decreases to zero. The Hsp70-mediated dissociation of the p58^{IPK}-Hdj1 complex is suggested to lead to an activation of p58^{IPK} allowing the interaction, monomerization and consequently inhibition of PKR (Melville et al. 1999). The exact mechanism of this process, in particular why two J-domain containing proteins are involved is not clear.

In summary, Hsp70 can be involved in viral gene expression at the level of transcription initiation and transcription elongation. In addition, Hsp70 is instumentalized by viruses to circumvent the general translation block induced by double-stranded RNA and interferon.

Morphogenesis

There is ample circumstantial evidence, based on "guilt by association", that Hsp70 systems may also be involved in viral morphogenesis assisting folding of capsid monomers, assembly of nucleocapsids, and facilitating folding of cytoplasmic of luminal domains of envelope proteins (Choukhi et al. 1998; Liberman et al. 1999; Macejak and Luftig 1991). Howerver, conclusive evidence that these interactions lead to higher yields of properly folded capsids or envelope proteins and more efficient virion assembly is still missing for most of the investigated viral model systems. A few more conclusive examples are detailed here. Hsp70 was shown to interact with the capsid proteins VP1, VP2, and VP3 of polyomavirus. Expression of these proteins in a variety of systems including A31 mouse fibroblasts, reticulocyte lysate, Sf9 insect cells, and *E. coli* leads to the formation of an ATP-sensitive complex with Hsp70 proteins. During infection the capsid protein–Hsp70 complex is first detected in the cytoplasm and subsequently imported into the nucleus. These observations prompted the speculation that Hsp70 assists folding of the capsid proteins to an assembly competent state but prevents premature virion assembly until translocation into the nucleus and genome replication have been completed (Cripe et al. 1995). Chromy et al. demonstrated that purified VP1 and VP3 assembles in vitro into polymorphic higher oligomeric structures upon addition of unphysiological concentrations of Ca^{2+} (0.5 mM), while the addition of the prokaryotic DnaK or mammalian Hsc70, which bound to the C terminus of VP1, inhibited the Ca^{2+} induced assembly. In contrast, the addition of the complete prokaryotic DnaK, DnaJ, GrpE chaperone team assembled VP1 and VP3 into virion-like structures in an ATP-dependent but Ca^{2+}-independent process. The mammalian Hsc70 could also assemble correct icosahedral virion particles in an ATP-dependent process when the SV40 large T antigen with a functional J-domain was present as its JDP partner (Chromy et al. 2003).

In the positive-stranded RNA closteroviruses Hsp70, which in this case is virus encoded as discussed in detail below, plays a different role in the assembly of the helical symmetric capsid. Genetic analysis demonstrated that deletion of the viral Hsp70, or mutations that abrogated its ATPase activity, dramatically reduced the formation of full-length virions (Satyanarayana et al. 2000). In a biochemical analysis of the filamentous virion particles the Hsp70 protein was found to be a component of the virion together with the major and the minor coat proteins (CP, CPm) and a fourth protein called p61 (Citrus tristeza virus, CTV) or p64 (Beet yellow virus) (Napuli et al. 2000, 2003; Satyanarayana et al. 2000). A more detailed analysis using a minimal CTV replicon, which contained only the gene encoding CPm with or without the genes encoding Hsp70 and p61, revealed that CPm starts incapsidation of the RNA at a 5′ nontranslated region, which previously was shown to be essential for virus replication (Gowda et al. 2003), and covers the RNA to different extents. When Hsp70 and p61 are present, incapsidation by CPm is restricted to about 630 nucleotides of the 5′ end consistent with the observation that in wild-type virions only a short tail is covered with CPm while the majority of the 20 kb RNA genome is covered with CP (Satyanarayana et al. 2004). Hsp70 therefore seems to be important for a coordinated incapsidation of the RNA. The mechanism of this process and whether Hsp70 has additional functions in the coat assembly reaction is unknown. It is also unclear why in contrast to other helical viruses like TMV two different coat proteins are necessary for the formation of this filamentous helical structure. One hypothesis is that CP forms a more stable coat around most of the RNA to protect the genome during the transition outside the plant cells, while encapsidation by CPm is less stable to allow efficient disassembly after reentry into a host cell. Thereby the origin of replication becomes accessible for translation and replication, two processes which could aid complete uncoating. The incorporation of Hsp70 into the CPm coated part of the capsid could stabilize CPm especially during extracellular transition where ATP is absent and Hsp70 release is slow.

During morphogenesis of the double-stranded RNA reovirus cellular Hsp70 assists the assembly of the trimeric lollipop-shaped sigma 1 protein that is responsible for the interaction with the host cell receptor. While the N-terminal filamentous part of the sigma 1 protein folds and trimerizes cotranslationally in a Hsp70-independent manner, the C-terminal globular domain folds post-translationally Hsp70-dependently. In this process Hsp70 binds already cotranslationally to a protein segment located downstream of the N-terminal triple α helical coiled-coil presumably inhibiting unwanted interactions and misfolding. After release from the ribosome trimerization of the C-terminal domain is coupled to ATP-mediated release of Hsp70 (Leone et al. 1996).

For the envelope proteins of a number of viruses including Sindbis virus, VSV, influenza A virus and HIV a transient interaction with the ER resident Hsp70 chaperone BiP/Grp78 was demonstrated (Carleton and Brown 1996; de Silva et al. 1990, 1993; Earl et al. 1991; Hammond and Helenius 1994; Hogue and Nayak 1992; Machamer et al. 1990; Mulvey and Brown 1995; Otteken et al. 1996; Singh et al. 1990). This interaction was prolonged when the folding of the protein or its assembly into an oligomeric structure was delayed by ATP depletion, prevention of disulfide bond formation, inhibition of glycosylation, temperature sensitive mutations at nonpermissive temperatures, and missing or misfolded oligomerization partners. These data strongly suggest that BiP plays an integral part of the folding of viral envelope proteins.

The cytoplasmic Hsp70 chaperone, however, can also be involved as demonstrated for the HBV glycoprotein L (Lambert and Prange 2003; Löffler-Mary et al. 1997; Prange et al. 1999). Hsc70, in conjunction with its J-domain co-chaperone Hdj1 and regulated in an antagonistic fashion by the co-chaperones Hip and Bag1, plays a functional role in the topogenesis of the L protein. The L protein is a polytopic membrane protein with initially three transmembrane helices which are inserted cotranslationally into the ER membrane in a topology where the N terminus is cytoplasmic and the C terminus is luminal. Post-translationally the N terminus is translocated into the ER lumen in approximately 50% of the molecules resulting in a mixed topology that is preserved during virus maturation allowing the N terminus to perform dual functions as nucleocapsid matrix protein and in receptor recognition (Bruss and Vieluf 1995; Le Seyec et al. 1999). Hsc70 binds to the N terminus and assists and regulates its translocation as evidenced by the following observations. Deletion of the Hsc70 binding site leads to a cotranslational translocation of the N terminus and a uniform topology (Löffler-Mary et al. 1997). Increasing intracellular Bag1 levels significantly enhance post-translational translocation of the N terminus consistent with its role as substrate release factor (see above; Gässler et al. 2001). In contrast, overproduction of Hip reduced post-translational translocation consistent with its proposed function as antagonist of Bag1 (Kanelakis et al. 2000). Together these data clearly show that the Hsp70 chaperone system regulates the mixed topology of the HBV envelope protein (Lambert and Prange 2003).

Taken together, Hsp70 assists folding and maturation of capsid and envelope proteins as well as the multimeric assembly reactions of subunits or entire virions.

Transformation

Viruses which do not provide their own polymerase are dependent on the host replication machinery. In order to proliferate in quiescent cells the virus has to reinitiate the cell cycle thereby transforming the cell. A number of mechanisms have evolved enabling viruses to overcome the restriction points of the cell cycle. The best investigated example demonstrating the involvement of Hsp70 systems is the DNA tumor virus SV40, the prototype of *Polyomaviridae* family, which also include the human BK and JC tumor viruses. Central to the transforming ability of SV40 virus are the large and small T antigen (TAg), both of which contain the signature motif of an Hsp70 co-chaperone, the J-domain, at their N terminus. Mutations within the J-domain that affect the functional interaction of the TAg with Hsp70s have been demonsrated to obliterate the ability of TAg to transform mammalian cells (Srinivasan et al. 1997) reviewed recently in full depth by Sullivan and Pipas (Sullivan and Pipas 2002). Among other functions the SV40 TAg sequesters the retinoblastoma family gene products pRb, p107 and p130 and liberates members of the E2F family of transcription factors in an Hsc70 and ATP hydrolysis-dependent manner (Sullivan et al. 2000a, 2001). The free E2F proteins subsequently trigger the expression of the S-phase genes leading to DNA replication (Dyson 1998). The most likely mechanism involves the following steps. First, the large TAg binds to the pRB–E2F complex. Second, Hsc70 in its ATP-bound form with high substrate association rates associates with the pRB–E2F–TAg complex. Third, the J-domain of TAg stimulates ATP hydrolysis by Hsc70 triggering the transition to the high-affinity conformation of Hsc70's substrate-binding domain thereby trapping the substrate protein, either pRB or E2F. Forth, Hsc70 induces a conformational change in the substrate protein leading to the dissociation of pRB and E2F, whereby pRB is still in complex with TAg and Hsc70 is bound to either pRB or E2F. ADP dissociation and rebinding of ATP to Hsc70 leads to the dissociation of the Hsc70–substrate complex resulting in free E2F and the pRB–TAg complex. The pRB–TAg complex may decay with a certain half-life spontaneously or induced through an ATPase driven interaction cycle with Hsc70 liberating TAg for other tasks. The fact that Hsc70 binds to TAg itself in an ATP-dependent manner is not surprising and also observed by other JDPs (e.g. Laufen et al. 1999; Mayer et al. 1999; Wittung-Stafshede et al.

2003). The J-domain signals to Hsp70 proteins the close proximity of a substrate. For lack of a substrate bound to the JDP Hsp70 binds to the JDP itself (compare Schirmbeck et al. 2002).

A second task of the TAg, the transactivation of E2F promoters independent of the disruption of the pRB–E2F complex also involves the J-domain and consequently a Hsp70 protein (Chao et al. 2000; Harris et al. 1998; Sheng et al. 1997; Sheng et al. 2000; Zalvide et al. 1998). A possible mechanism could involve the Hsc70 assisted assembly of a transcription initiation complex on the respective promoters similar to the role of Hsc70 in the Bag1 stimulated transcription discussed above. Alternatively, TAg could induce Hsc70 to disassemble an inhibitory silencer complex or to assist in remodeling the chromatin.

Although no clear indication was found so far for a participation of the J-domain of TAg in the interaction with p53 and ensuing inactivation of p53 which contributes to cell transformation and prevention of apoptosis, it should be mentioned that p53 interacts with the Hsp70–Hsp90 chaperone machinery (King et al. 2001; Zylicz et al. 2001) and an Hsp70-mediated transfer of p53 into a complex with TAg is possible. A JDP involved in this process may be TPR2 instead of TAg because it was suggested that TPR2 mediates the retrograde transfer of an Hsp90 bound client onto Hsp70 (Brychzy et al. 2003).

HPV and AdVs have similar transforming activities using the proteins E7 (HPV) and E1A (AdV) to disrupt the complexes between pRB family proteins and E2F family proteins. Although neither E7 nor E1A contains a J-domain both proteins could transform cells in a way similar to that described for SV40 TAg. E7 interacts with the JDP hTid-1, the homolog of the *Drosophila* protein Tid56, which acts as a tumor suppressor (Schilling et al. 1998). The C terminus of E7, which mediates the interaction with hTid-1, has also been shown to be essential for the physical disruption of the pRB–E2F complex albeit it is not necessary for direct interaction with pRB (Patrick et al. 1994; Wu et al. 1993). These observations suggest that the interaction with hTid-1 is involved in the disruption of the pRB–E2F complex providing E7 with the J-domain necessary to recruit Hsc70 for the complex dissociation in analogy to the function of SV40 TAg. Alternatively, the binding of E7 to hTid-1 could transform cells through an inhibition of the assumed tumor suppressor function of hTid-1. E1A, on the other side, directly interacts with Hsc70 (White et al. 1988) and could use Hsc70 in this way to disrupt the pRB–E2F complex. However, in addition to Hsc70 a JDP would be necessary for efficient stimulation of Hsc70's ATPase activity.

In conclusion it can be said that most double-stranded DNA viruses depend on Hsp70 chaperones for reprogramming of the host cells to reenter into the cell cycle. The dependency of cell transformation on Hsp70 chaperones is also observed in many tumor cells.

Cell survival and apoptosis

Hsp70 systems are essential for the survival under stress conditions. Therefore, the induction of Hsp70 protein production may serve an additional purpose for the virus, namely, the prevention or delay of host cell death until the progeny is ready to leave. It is therefore not surprising that interference with apoptotic signal transduction pathways in both directions–inhibiting and activating–is a quite common phenomenon accompanying viral infections (reviewed in Benedict et al. 2002; Hardwick 2001; Roulston et al. 1999). The induction of cell death during viral infection can have different reasons. First, many DNA viruses replicating in quiescent cells need to induce reentry into the cell cycle by providing proliferation signals. Such untimely proliferation signals very often induce apoptosis as first line of cellular defense against viruses as well as against tumor growth (Debbas and White 1993; Fromm et al. 1994; Lowe and Ruley 1993; Symonds et al. 1994). Second, many different kinds of stress induce apoptosis and viral infection may impose such a lethal stress. Third, viral infection of a cell can induce the release of interferons triggering apoptosis (Castelli et al. 1997; Diaz-Guerra et al. 1997; Rivas et al. 1998). It is self-evident that viruses have evolved means to interfere with the cell death response because apoptosis is detrimental for viral proliferation during the early phase of viral life cycle. In contrast, when the viral particles are ready assembled apoptosis can be advantageous for virus spreading (Anderson 2001; Best et al. 2002; Goh et al. 2001; Henke et al. 2000, 2001; Muthumani et al. 2002a, 2002b; Schultz-Cherry et al. 2001). If virions are incorporated into apoptotic bodies which are engulfed by neighboring phagocytic cells, viruses can infect new cells without the danger of being exposed to the host's immune system (Fazakerley 2001; Fazakerley and Allsopp 2001).

It has been shown that Hsp70 plays an important role in apoptosis. Hsp70 interferes with the signal transduction pathways leading to programmed cell death at several levels including inhibition of (1) the apoptosis inducing Jun N-terminal kinase, (2) cytocrome C release out of mitochondria, (3) apoptosome assembly by interaction with Apaf-1, and (4) execution of apoptosis acting even downstream of caspase 3 (Beere and Green 2001; Beere et al. 2000; Gabai et al. 1997, 1998; Jäättelä et al. 1998; Jolly and Morimoto 2000; Li et al. 2000; Mosser et al. 1997, 2000; Ravagnan et al. 2001; Saleh et al. 2000; Yaglom et al. 1999). In addition, Jäättelä and coworkers demonstrated that depletion of Hsp70 in *hsp70* overexpressing tumor cells leads to an induction of apoptosis (Nylandsted et al. 2000). It is therefore conceivable that upregulation of *hsp70* expression may also serve as a tool to prevent premature apoptotic cell death, and the precipitous decrease of *hsp70* mRNA levels at the end of AdV infections may lead to the timed induction of apoptosis.

In this context it is also interesting that a decrease of the BiP concentration in tumor cell lines inhibits tumor progression and eliminates resistance to T-cell mediated cytotoxicity (Jamora et al. 1996; Sugawara et al. 1993). In analogy, viruses may evade T-cell mediated induction of apoptosis by upregulating the concentration of the ER resident Hsp70 homolog.

Virus-encoded components of the Hsp70 system

Not all viruses seem to be content with inducing the expression of the genes encoding the components of the Hsp70 system and then compete with cellular substrates for their chaperone power. There are examples of viruses that bring with them their own components of the Hsp70 system.

Viral J-domain containing proteins

The multifunctional large and small TAg of the simian and human DNA tumor viruses SV40, JC and BK virus, contain an N-terminal J-domain that functionally interacts with Hsp70 chaperones acting like a bona fide J-domain (Campbell et al. 1997; Genevaux et al. 2003; Kelley and Georgopoulos 1997; Kelley and Landry 1994). This J-domain induces Hsc70 binding to the TAg in an ATP-dependent manner (Sullivan et al. 2001). The TAg are involved in almost all viral activities, including genome replication, regulation of transcription of viral and cellular genes, virion assembly, and cell transformation and many of these functions are abrogated by mutations in the J-domain (DeCaprio 1999; Kelley 1999). It is therefore assumed that its mode of action is based on the targeting of the Hsp70 chaperone machine to certain protein complexes inducing remodeling and change of activity of these complexes (for review see Sullivan and Pipas 2002).

The MC013L protein of the DNA poxvirus causing molluscum contagiosum also contains a J-domain (Moratilla et al. 1997). This protein interacts with the glucocorticoid and vitamin D receptors and downregulates their transcriptional activity, possibly by interfering with the Hsp70–Hsp90 chaperone machine (Chen et al. 2000).

The need for a specialized chaperone to assist folding and assembly of viral and host proteins is reminiscent of the T4-like phages that encode an Hsp60 co-chaperone in order to ensure the proper folding of their capsid protein (for a review see Ang et al. 2000).

The question arises why the polyoma- and poxviruses bring their own JDP along. Two alternative hypotheses are possible. First, the viral J-domain has a unique property that is not present in cellular JDPs. Second, the fusion of a J-domain to a viral protein could be more beneficial for the protein's function than the recruitment of a host JDP would be. Comparing the NMR structures of the J-domain of murin TAg and human Hdj1 reveals clear differences, in particular, when the electrostatic potential at the surface is analyzed (Fig. 4). In addition, many of the residues corresponding to *E. coli* DnaJ residues which are important for interaction with DnaJ's Hsp70 partner DnaK are conserved in Hdj1 but not in TAg (Fig. 4) (Genevaux et al. 2002; Greene et al. 1998). Nevertheless, the J-domains of SV40, JC and BK TAg can substitute for the J-domain of DnaJ in supporting Hsp70 function in *E. coli* (Genevaux et al. 2003; Kelley and Georgopoulos 1997). To address the question of specific effects of the TAg J-domain in more detail, domain-swap experiments were carried out and TAg function was investigated. The J-domain of SV40 TAg was replaced by two different J-domains of cellular JDPs without loss of functionality as measured by the ability to promote viral DNA replication (Campbell et al. 1997), interaction with pRB family proteins, and in vivo viral proliferation (Stubdal et al. 1997). These results clearly suggest that the viral J-domain functions as a normal J-domain without any specific properties. On the other hand, the J-domain of polyomaviruses contains in helix 1 a highly conserved ^{13}LXXLLXL^{19} motif that is not conserved in other J-domains. Mutational alteration of this sequence reduced viral DNA replication 100-fold while the effects on pRB, including release and activation of E2F, were unimpaired (Li et al. 2001). Magnusson and coworkers suggest from these data that the sequence of the viral J-domain may have, in some contexts, a specific function. However, of the three substitutions tested (L13M, L13I, L13V) only L13V had a phenotype. Given the fact that the consensus residue in other JDPs is the large hydrophobic amino acid tyrosine it is likely that a minimum size is necessary to stabilize the hydrophobic core between helix 1, 2, and 3 and valine is just too small, leading to a destabilization of helix 1 or the entire J-domain. This interpretation is supported by the fact that the small TAg with this mutation is unstable.

Taken together, all available evidence indicates that the evolutionary fusion of a J-domain to the viral protein increased the functionality of this protein by the more efficient recruitment of the Hsp70 chaperones as compared to other viral proteins that interact with host JDPs.

Virus-encoded Hsp70 proteins

The plant pathogenic closteroviruses, including beet yellow virus, lettuce infectious yellow virus, and CTV, went one step further and encode for their own proper Hsp70 chaperone. This Hsp70 protein is essential for cell-to-cell transmission (Agranovsky et al. 1998; Alzhanova et al. 2000, 2001; Peremyslov et al. 1999) and has been implicated in virion assembly as discussed above (Napuli et al. 2003; Napuli et al. 2000; Satyanarayana et al. 2000, 2004). In addition to its function in virion assembly, the viral Hsp70 also binds to microtubules and is found associated with the plasmodesmata, the open cytoplasmic cell-to-cell contacts, in closterovirus infected plant cells (Karasev et al. 1992; Medina et al. 1999) and interacts with a long-distance transport factor (Prokhnevsky et al. 2002). It is therefore possible that it has supplementary functions in viral transmission.

Similar to the situation with the virus-encoded JDPs, there are two possible explanations why closteroviruses encode for their own Hsp70 protein. First, coding for and regulating the production of a proper Hsp70 chaperone could supplement the viral need for chaperone power and even make the virus independent of the host Hsp70s. Second, the virus-encoded Hsp70 could have specific properties of the ATPase cycle and substrate specificity/affinity which are necessary for the folding of specific viral and host substrates but not supplied by the cellular Hsp70 proteins. Since it has not yet been attempted to complement the loss of function of the viral Hsp70 by the overproduction of a host Hsp70 protein, and the viral Hsp70s are not yet biochemically characterized, it is not possible to distinguish the two alternative hypotheses at present. However, in the unrelated tomato spotted wilt virus, a negative-stranded RNA virus of the *Bunyaviridae* family, the viral movement protein binds to the viral nucleocapsid and a plant JDP indicating the recruitment of a host Hsp70 for the cell-to-cell movement process (Soellick et al. 2000). This observation demonstrates that the plant virus specific process of cell-to-cell movement as such does not require a specialized Hsp70 chaperone. Furthermore, the fact that the Hsp70 proteins of the different members of the *Closteroviridae* family are not highly conserved (7.6% identity, 25% similarity in a multiple sequence alignment; see Fig. 2) and only slightly closer related among each other than to human Hsc70 (87% of the residues completely conserved within the *Closteroviridae* Hsp70s are also found in human Hsc70) does not support the hypothesis of an adaptation to a specific folding task. However, the co-evolution of the viral Hsp70 with a specific viral protein such as the small capsid protein would not necessarily be apparent in sequence identity among the different *Closteroviridae* Hsp70s. Furthermore, all viral-encoded Hsp70s are on average 82 residues shorter than the eukaryotic cytosolic Hsp70 proteins. Since the viral Hsp70s remain incorporated in the nucleocapsid (about 10 molecules per capsid; Napuli et al. 2000), the size of the protein may be an issue.

Taken together, although the currently available evidence provides no indication for an adaptation to a specific folding task, co-evolution with a specific substrate (capsid protein) and the requirement to fit into the nucleocapsid cannot be excluded as a driving force for the evolution of a virus-encoded Hsp70 protein.

Evolutionary aspects of the virus–Hsp70 interaction

Many mutational alterations lead to a destabilization of proteins and increase their tendency to misfold and aggregate. Since chaperones recognize misfolded proteins, prevent their aggregation and promote their refolding into the native state, chaperones can keep a mutationally altered protein functional and thereby buffer detrimental mutations. Such a function as evolutionary capacitor was shown for the Hsp90 chaperone (Queitsch et al. 2002; Rutherford and Lindquist 1998; Sollars et al. 2003). Rutherford and Lindquist demonstrated that the reduction of Hsp90 activity in *Drosophila melanogaster* by genetic or pharmacological means leads to the appearance of strain specific morphological alterations suggesting that Hsp90 preserves the natural function of certain morphogenetic factors that in its absence would not be active or would have a different activity. Continuous selection for a given trait under low Hsp90 activity conditions (mutant Hsp90, presence of an Hsp90 inhibitor or stress) could make the trait independent of Hsp90 activity. Similarly, Queitsch and Lindquist showed that plasticity in plants is also Hsp90 dependent. In principle, a similar function could also be performed by Hsp70 chaperones. Like Hsp90, Hsp70s are involved in chaperoning signaling molecules such as transcription factors and protein kinases, which constitute some of the morphogenetic factors. Since Hsp70, but not Hsp90, can refold denatured proteins Hsp70s could buffer many destabilizing alterations in structual proteins.

RNA replication is a very error-prone process with a rate of 1 misincorporation per 10^4 replicated bases. In the absence of repair mechanisms this means that the progeny viruses differ from each other and from the parent, a fact that is expressed in the term ‘quasi-species’ for RNA viruses because they are defined by a sequence space and not by a unique sequence found in the majority of the species’ members. The virus-encoded Hsp70 proteins are an impressive example for the accumulation of mutational alterations. While Hsp70s belong to the most conserved proteins with about 50% sequence identity between *E. coli* and man, the sequence variations found within the virus-encoded Hsp70s is so high that only 7.6% of the residues remained unchanged within the entire family (see Fig. 2). It therefore is very likely that destabilizing mutations occur frequently in viral genes increasing the likelihood for chaperone dependency of the encoded proteins. The minimal number of fully functional progeny required to give a sufficiently high chance for a successful new infection defines the mutational window for the virus species which is widened by the action of Hsp70 chaperones. Under certain circumstances Hsp70 chaperones may even be the prerequisite for an evolutionary survival of RNA viruses.

Concluding remarks

The ample evidence demonstrating the participation of Hsp70 chaperones at distinct steps in the life cycles of a variety of viruses suggests that Hsp70 involvement in virus proliferation is a widespread phenomenon. Since Hsp70 function is essential for eukaryotic cells and complete knockouts of the major *hsp70* genes are therefore not available, it is difficult to show in how many viral processes Hsp70s are involved and to rigorously demonstrate that the Hsp70 contribution to these processes is essential. The reduction of the Hsp70 levels through the use of gene knockdown technologies (siRNA) may give valuable indications in future studies. In addition, as more and more in vitro reconstitution systems for the different steps of the viral life cycle become available, the mechanism of the Hsp70-mediated viral folding and assembly processes will be elucidated.

Acknowledgements. I thank Drs. R. Wegrzyn and K. Turgay for helpful comments on the manuscript.

References

Agoff SN, Wu B (1994) CBF mediates adenovirus Ela trans-activation by interaction at the C-terminal promoter targeting domain of conserved region 3. Oncogene 9:3707–3711

Agostini I, Popov S, Li J, Dubrovsky L, Hao T, Bukrinsky M (2000) Heat-shock protein 70 can replace viral protein R of HIV-1 during nuclear import of the viral preintegration complex. Exp Cell Res 259:398–403

Agranovsky AA, Folimonov AS, Folimonova S, Morozov S, Schiemann J, Lesemann D, Atabekov JG (1998) Beet yellows closterovirus HSP70-like protein mediates the cell-to-cell movement of a potexvirus transport-deficient mutant and a hordeivirus-based chimeric virus. J Gen Virol 79:889–895

Alfano C, McMacken R (1989a) Heat shock protein-mediated disassembly of nucleoprotein structures is required for the initiation of bacteriophage lambda DNA replication. J Biol Chem 264:10709–10718

Alfano C, McMacken R (1989b) Ordered assembly of nucleoprotein structures at the bacteriophage lambda replication origin during the initiation of DNA replication. J Biol Chem 264:10699–10708

Alzhanova DV, Hagiwara Y, Peremyslov VV, Dolja VV (2000) Genetic analysis of the cell-to-cell movement of beet yellows closterovirus. Virology 268:192–200

Alzhanova DV, Napuli AJ, Creamer R, Dolja VV (2001) Cell-to-cell movement and assembly of a plant closterovirus: roles for the capsid proteins and Hsp70 homolog. EMBO J 20:6997–7007

Anderson JR (2001) The mechanisms of direct, virus-induced destruction of neurons. Curr Top Microbiol Immunol 253:15–33

Ang D, Keppel F, Klein G, Richardson A, Georgopoulos C (2000) Genetic analysis of bacteriophage-encoded cochaperonins. Annu Rev Genet 34:439–456

Arias CF, Isa P, Guerrero CA, Mendez E, Zarate S, Lopez T, Espinosa R, Romero P, Lopez S (2002) Molecular biology of rotavirus cell entry. Arch Med Res 33:356–361

Arnold-Schild D, Hanau D, Spehner D, Schmid C, Rammensee HG, de la Salle H, Schild H (1999) Cutting edge: receptor-mediated endocytosis of heat shock proteins by professional antigen-presenting cells. J Immunol 162:3757–3760

Arsene F, Tomoyasu T, Bukau B (2000) The heat shock response of *Escherichia coli*. Int J Food Microbiol 55:3–9

Asea A (2003) Chaperokine-induced signal transduction pathways. Exerc Immunol Rev 9:25–33

Asea A, Rehli M, Kabingu E, Boch JA, Bare O, Auron PE, Stevenson MA, Calderwood SK (2002) Novel signal transduction pathway utilized by extracellular HSP70: role of toll-like receptor (TLR) 2 and TLR4. J Biol Chem 277:15028–15034

Ballinger CA, Connell P, Wu Y, Hu Z, Thompson LJ, Yin LY, Patterson C (1999) Identification of CHIP, a novel tetratricopeptide repeat-containing protein that interacts with heat shock proteins and negatively regulates chaperone functions. Mol Cell Biol 19:4535–4545

Bardelli A, Longati P, Alberto D, Goruppi S, Schneider C, Ponzetto C, Comoglio PM (1996) HGF receptor associates with the anti-apoptotic protein BAG-1 and prevents cell death. EMBO J. 15:6205–6212

Barouch W, Prasad K, Greene L, Eisenberg E (1997) Auxilin-induced interaction of the molecular chaperone Hsc70 with clathrin baskets. Biochemistry 36:4303–4308

Bartenschlager R, Junker-Niepmann M, Schaller H (1990) The P gene product of hepatitis B virus is required as a structural component for genomic RNA encapsidation. J Virol 64:5324–5332

Bartenschlager R, Schaller H (1992) Hepadnaviral assembly is initiated by polymerase binding to the encapsidation signal in the viral RNA genome. Embo J 11:3413–3420

Beck J, Nassal M (2001) Reconstitution of a functional duck hepatitis B virus replication initiation complex from separate reverse transcriptase domains expressed in *Escherichia coli*. J Virol 75:7410–7419

Beck J, Nassal M (2003) Efficient Hsp90-independent in vitro activation by Hsc70 and Hsp40 of duck hepatitis B virus reverse transcriptase, an assumed Hsp90 client protein. J Biol Chem 278:36128–36138

Becker T, Hartl FU, Wieland F (2002) CD40, an extracellular receptor for binding and uptake of Hsp70-peptide complexes. J Cell Biol 158:1277–1285

Beere HM, Green DR (2001) Stress management—heat shock protein-70 and the regulation of apoptosis. Trends Cell Biol 11:6–10

Beere HM, Wolf BB, Cain K, Mosser DD, Mahboubi A, Kuwana T, Tailor P, Morimoto RI, Cohen GM, Green DR (2000) Heat-shock protein 70 inhibits apoptosis by preventing recruitment of procaspase-9 to the Apaf-1 apoptosome. Nat Cell Biol. 2:469–475

Benedict CA, Norris PS, Ware CF (2002) To kill or be killed: viral evasion of apoptosis. Nat Immunol 3:1013–1018

Berjanskii MV, Riley MI, Xie A, Semenchenko V, Folk WR, Van Doren SR (2000) NMR structure of the N-terminal J domain of murine polyomavirus T antigens. Implications for DnaJ-like domains and for mutations of T antigens. J Biol Chem 275:36094–36103

Best SM, Wolfinbarger JB, Bloom ME (2002) Caspase activation is required for permissive replication of Aleutian mink disease parvovirus in vitro. Virology 292:224–234

Bimston D, Song J, Winchester D, Takayama S, Reed JC, Morimoto RI (1998) BAG-1, a negative regulator of Hsp70 chaperone activity, uncouples nucleotide hydrolysis from substrate release. EMBO J 17:6871–6878

Binder RJ, Harris ML, Menoret A, Srivastava PK (2000) Saturation, competition, and specificity in interaction of heat shock proteins (hsp) gp96, hsp90, and hsp70 with CD11b+ cells. J Immunol 165:2582–2587

Blond-Elguindi S, Cwirla SE, Dower WJ, Lipshutz RJ, Sprang SR, Sambrook JF, Gething MJ (1993) Affinity panning of a library of peptides displayed on bacteriophages reveals the binding specificity of BiP. Cell 75:717–728

Bolt G (2001) The measles virus (MV) glycoproteins interact with cellular chaperones in the endoplasmic reticulum and MV infection upregulates chaperone expression. Arch Virol 146:2055–2068

Boulanger J, Faulds D, Eddy EM, Lingwood CA (1995) Members of the 70 kDa heat shock protein family specifically recognize sulfoglycolipids: role in gamete recognition and mycoplasma-related infertility. J Cell Physiol 165:7–17

Brehmer D, Rüdiger S, Gässler CS, Klostermeier D, Packschies L, Reinstein J, Mayer MP, Bukau B (2001) Tuning of chaperone activity of Hsp70 proteins by modulation of nucleotide exchange. Nat Struct Biol 8:427–432

Brenner BG, Wainberg MA (1999) Heat shock protein-based therapeutic strategies against human immunodeficiency virus type 1 infection. Infect Dis Obstet Gynecol 7:80–90

Briknarova K, Takayama S, Brive L, Havert ML, Knee DA, Velasco J, Homma S, Cabezas E, Stuart J, Hoyt DW, Satterthwait AC, Llinas M, Reed JC, Ely KR (2001) Structural analysis of BAG1 cochaperone and its interactions with Hsc70 heat shock protein. Nat Struct Biol 8:349–352

Brum LM, Lopez MC, Varela JC, Baker HV, Moyer RW (2003) Microarray analysis of A549 cells infected with rabbitpox virus (RPV): a comparison of wild-type RPV and RPV deleted for the host range gene, SPI-1. Virology 315:322–334

Bruss V, Vieluf K (1995) Functions of the internal pre-S domain of the large surface protein in hepatitis B virus particle morphogenesis. J Virol 69:6652–6657

Brychzy A, Rein T, Winklhofer KF, Hartl FU, Young JC, Obermann WM (2003) Cofactor Tpr2 combines two TPR domains and a J domain to regulate the Hsp70/Hsp90 chaperone system. EMBO J 22:3613–3623

Buchner E, Gundersen CB (1997) The DnaJ-like cysteine string protein and exocytotic neurotransmitter release. Trends Neurosci 20:223–227

Bukau B, Deuerling E, Pfund C, Craig EA (2000) Getting newly synthesized proteins into shape. Cell 101:119–122

Campbell KS, Mullane KP, Aksoy IA, Stubdal H, Zalvide J, Pipas JM, Silver PA, Roberts TM, Schaffhausen BS, DeCaprio JA (1997) DnaJ/hsp40 chaperone domain of SV40 large T antigen promotes efficient viral DNA replication. Genes Dev 11:1098–1110

Carleton M, Brown DT (1996) Disulfide bridge-mediated folding of Sindbis virus glycoproteins. J Virol 70:5541–5547

Carleton M, Brown DT (1997) The formation of intramolecular disulfide bridges is required for induction of the Sindbis virus mutant ts23 phenotype. J Virol 71:7696–7703

Castelli JC, Hassel BA, Wood KA, Li XL, Amemiya K, Dalakas MC, Torrence PF, Youle RJ (1997) A study of the interferon antiviral mechanism: apoptosis activation by the 2–5A system. J Exp Med 186:967–972

Caswell R, Hagemeier C, Chiou CJ, Hayward G, Kouzarides T, Sinclair J (1993) The human cytomegalovirus 86 K immediate early (IE) 2 protein requires the basic region of the TATA-box binding protein (TBP) for binding, and interacts with TBP and transcription factor TFIIB via regions of IE2 equired for transcriptional regulation. J Gen Virol 74:2691–2698

Chao HH, Buchmann AM, DeCaprio JA (2000) Loss of p19(ARF) eliminates the requirement for the pRB-binding motif in simian virus 40 large T antigen-mediated transformation. Mol Cell Biol 20:7624–7633

Cheetham MIE, Caplan AJ (1998) Structure, function and evolution of DnaJ: conservation and adaptation of chaperone function. Cell Stress Chap 3:28–36

Chen D, Fong Y, Zhou Q (1999) Specific interaction of Tat with the human but not rodent P-TEFb complex mediates the species-specific Tat activation of HIV-1 transcription. Proc Natl Acad Sci USA 96:2728–2733

Chen N, Baudino T, MacDonald PN, Green M, Kelley WL, Burnett JW, Buller RM (2000) Selective inhibition of nuclear steroid receptor function by a protein from a human tumorigenic poxvirus. Virology 274:17–25

Cheng H, Cenciarelli C, Shao Z, Vidal M, Parks WP, Pagano M, Cheng-Mayer C (2001) Human T cell leukemia virus type 1 Tax associates with a molecular chaperone complex containing hTid-1 and Hsp70. Curr Biol 11:1771–1775

Cheung RK, Dosch HM (1993) The growth transformation of human B cells involves superinduction of hsp70 and hsp90. Virology 193:700–708

Cho DY, Yang GH, Ryu CJ, Hong HJ (2003) Molecular chaperone GRP78/BiP interacts with the large surface protein of hepatitis B virus in vitro and in vivo. J Virol 77:2784–2788

Chouchane L, Bowers FS, Sawasdikosol S, Simpson RM, Kindt TJ (1994) Heat-shock proteins expressed on the surface of human T cell leukemia virus type I-infected cell lines induce autoantibodies in rabbits. J Infect Dis 169:253–259

Choukhi A, Ung S, Wychowski C, Dubuisson J (1998) Involvement of endoplasmic reticulum chaperones in the folding of hepatitis C virus glycoproteins. J Virol 72:3851–3818

Chroboczek J, Gout E, Favier AL, Galinier R (2003) Novel partner proteins of adenovirus penton. Curr Top Microbiol Immunol 272:37–55

Chromy LR, Pipas JM, Garcea RL (2003) Chaperone-mediated in vitro assembly of Polyomavirus capsids. Proc Natl Acad Sci USA 100:10477–10482

Clapham P, Nagy K, Cheingsong-Popov R, Exley M, Weiss RA (1983) Productive infection and cell-free transmission of human T-cell leukemia virus in a nonlymphoid cell line. Science 222:1125–1127

Clevenger CV, Thickman K, Ngo W, Chang WP, Takayama S, Reed JC (1997) Role of Bag-1 in the survival and proliferation of the cytokine-dependent lymphocyte lines, Ba/F3 ad Nb2. Mol Endocrinol 11:608–618

Colberg-Poley AM, Santomenna LD (1988) Selective induction of chromosomal gene expression by human cytomegalovirus. Virology 166:217–228

Colberg-Poley AM, Santomenna LD, Harlow PP, Benfield PA, Tenney DJ (1992) Human cytomegalovirus US3 and UL36–38 immediate-early proteins regulate gene expression. J Virol 66:95–105

Collins PL, Hightower LE (1982) Newcastle disease virus stimulates the cellular accumulation of stress (heat shock) mRNAs and proteins. J Virol 44:703–707

Connell P, Ballinger CA, Jiang J, Wu Y, Thompson LJ, Hohfeld J, Patterson C (2001) The co-chaperone CHIP regulates protein triage decisions mediated by heat-shock proteins. Nat Cell Biol 3:93–96

Corsi AK, Schekman R (1996) Mechanism of polypeptide translocation into the endoplasmic reticulum. J Biol Chem 271:30299–30302

Craig E, Yan W, James P (1999) Genetic dissection of the Hsp70 chaperone system of yeast. In: Bukau B (ed) Molecular Chaperones and Folding Catalysts. Regulation, Cellular Function and Mechanisms. Harwood Academic Publishers, Amsterdam, pp 139–162

Cripe TP, Delos SE, Estes PA, Garcea RL (1995) In vivo and in vitro association of hsc70 with polyomavirus capsid proteins. J Virol 69:7807–7813

Cullen BR (1998) HIV-1 auxiliary proteins: making connections in a dying cell. Cell 93:685–692

D'Silva PD, Schilke B, Walter W, Andrew A, Craig EA (2003) J protein cochaperone of the mitochondrial inner membrane required for protein import into the mitochondrial matrix. Proc Natl Acad Sci USA 100:13839–13844

Dai Q, Zhang C, Wu Y, McDonough H, Whaley RA, Godfrey V, Li H-H, Madamanchi N, Xu W, Neckers L, Cyr DM, Patterson C (2003) CHIP activates HSF1 and confers protection against apoptosis and cellular stress. EMBO J 22:5446–5458

Damania B, Lieberman P, Alwine JC (1998a) Simian virus 40 large T antigen stabilizes the TATA-binding protein- TFIIA complex on the TATA element. Mol Cell Biol 18:3926–3935

Damania B, Mital R, Alwine JC (1998b) Simian virus 40 large T antigen interacts with human TFIIB-related factor and small nuclear RNA-activating protein complex for transcriptional activation of TATA-containing polymerase III promoters. Mol Cell Biol 18:1331–1338 de Silva A, Braakman I, Helenius A (1993) Posttranslational folding of vesicular stomatitis virus G protein in the ER: involvement of noncovalent and covalent complexes. J Cell Biol 120:647–655

de Silva AM, Balch WE, Helenius A (1990) Quality control in the endoplasmic reticulum: folding and misfolding of vesicular stomatitis virus G protein in cells and in vitro. J Cell Biol 111:857–866

Debbas M, White E (1993) Wild-type p53 mediates apoptosis by E1A, which is inhibited by E1B. Genes Dev 7:546–554

DeCaprio JA (1999) The role of the J domain of SV40 large T in cellular transformation. Biologicals 27:23–28

Devireddy LR, Kumar KU, Pater MM, Pater A (2000) BAG-1, a novel Bcl-2-interacting protein, activates expression of human JC virus. J Gen Virol 81:351–357

Diaz-Guerra M, Rivas C, Esteban M (1997) Activation of the IFN-inducible enzyme RNase L causes apoptosis of animal cells. Virology 236:354–363

Dibbens JA, Muraiso K, Chattoraj DK (1997) Chaperone-mediated reduction of RepA dimerization is associated with RepA conformational change. Mol Microbiol 26:185–195

Dodson M, Echols H, Wickner S, Alfano C, Mensa-Wilmot K, Gomes B, LeBowitz J, Roberts JD, McMacken R (1986) Specialized nucleoprotein structures at the origin of replication of bacteriophage lambda: localized unwinding of duplex DNA by a six-protein reaction. Proc Natl Acad Sci USA 83:7638–7642

Dodson M, McMacken R, Echols H (1989) Specialized nucleoprotein structures at the origin of replication of bacteriophage lambda. Protein association and disassociation reactions responsible for localized initiation of replication. J Biol Chem 264:10719–10725

Dodson M, Roberts J, McMacken R, Echols H (1985) Specialized nucleoprotein structures at the origin of replication of bacteriophage lambda: complexes with lambda O protein and with lambda O, lambda P, and *Escherichia coli* DnaB proteins. Proc Natl Acad Sci USA 82:4678–4682

Domanico SZ, DeNagel DC, Dahlseid JN, Green JM, Pierce SK (1993) Cloning of the gene encoding peptide-binding protein 74 shows that it is a new member of the heat shock protein 70 family. Mol Cell Biol 13:3598–3610

Dornreiter I, Hoss A, Arthur AK, Fanning E (1990) SV40 T antigen binds directly to the large subunit of purified DNA polymerase alpha. EMBO J 9:3329–3336

Drahos DJ, Hendrix RW (1982) Effect of bacteriophage lambda infection on synthesis of groE protein and other *Escherichia coli* proteins. J Bacteriol 149:1050–1063

Dybdahl B, Wahba A, Lien E, Flo TH, Waage A, Qureshi N, Sellevold OF, Espevik T, Sundan A (2002) Inflammatory response after open heart surgery: release of heat-shock protein 70 and signaling through toll-like receptor-4. Circulation 105:685–690

Dyson N (1998) The regulation of E2F by pRB-family proteins. Genes Dev 12:2245–2262

Earl PL, Moss B, Doms RW (1991) Folding, interaction with GRP78-BiP, assembly, and transport of the human immunodeficiency virus type 1 envelope protein. J Virol 65:2047–2055

Eom CY, Lehman IR (2002) The human DnaJ protein, hTid-1, enhances binding of a multimer of the herpes simplex virus type 1 UL9 protein to oris, an origin of viral DNA replication. Proc Natl Acad Sci USA 99:1894–1898

Evans GJ, Morgan A, Burgoyne RD (2003) Tying everything together: the multiple roles of cysteine string protein (CSP) in regulated exocytosis. Traffic 4:653–659

Fang D, Haraguchi Y, Jinno A, Soda Y, Shimizu N, Hoshino H (1999) Heat shock cognate protein 70 is a cell fusion-enhancing factor but not an entry factor for human T-cell lymphotropic virus type I. Biochem Biophys Res Commun 261:357–363

Fazakerley JK (2001) Neurovirology and developmental neurobiology. Adv Virus Res 56:73–124

Fazakerley JK, Allsopp TE (2001) Programmed cell death in virus infections of the nervous system. Curr Top Microbiol Immunol 253:95–119

Fromm L, Shawlot W, Gunning K, Butel JS, Overbeek PA (1994) The retinoblastoma protein-binding region of simian virus 40 large T antigen alters cell cycle regulation in lenses of transgenic mice. Mol Cell Biol 14:6743–6754

Furlini G, Vignoli M, Re MC, Gibellini D, Ramazzotti E, Zauli G, La Placa M (1994) Human immunodeficiency virus type 1 interaction with the membrane of CD4+ cells induces the synthesis and nuclear translocation of 70 K heat shock protein. J Gen Virol 75:193–199

Gabai VL, Meriin AB, Mosser DD, Caron AW, Rits S, Shifrin VI, Sherman MY (1997) Hsp70 prevents activation of stress kinases. A novel pathway of cellular thermotolerance. J Biol Chem 272:18033–18037

Gabai VL, Meriin AB, Yaglom JA, Volloch VZ, Sherman MY (1998) Role of Hsp70 in regulation of stress-kinase JNK: implications in apoptosis and aging. FEBS Lett 438:1–4

Galabru J, Hovanessian A (1987) Autophosphorylation of the protein kinase dependent on double-stranded RNA. J Biol Chem 262:15538–15544

Gale M, Jr., Blakely CM, Hopkins DA, Melville MW, Wambach M, Romano PR, Katze MG (1998) Regulation of interferon-induced protein kinase PKR: modulation of P58IPK inhibitory function by a novel protein, P52rIPK. Mol Cell Biol 18:859–871

Gässler CS, Wiederkehr T, Brehmer D, Bukau B, Mayer MP (2001) Bag-1 M accelerates nucleotide release for human Hsc70 and Hsp70 and can act concentration-dependent as positive and negative cofactor. J Biol Chem 276:32538–32544

Genevaux P, Lang F, Schwager F, Vartikar JV, Rundell K, Pipas JM, Georgopoulos C, Kelley WL (2003) Simian virus 40 T antigens and J domains: analysis of Hsp40 cochaperone functions in *Escherichia coli*. J Virol 77:10706–10713

Genevaux P, Schwager F, Georgopoulos C, Kelley WL (2002) Scanning mutagenesis identifies amino acid residues essential for the in vivo activity of the *Escherichia coli* DnaJ (Hsp40) J-domain. Genetics 162:1045–1053

Georgopoulos CP (1977) A new bacterial gene (*groPC*) which affects lambda DNA replication. Mol Gen Genet 151:35–39

Georgopoulos CP, Hendrix RW, Kaiser AD, Wood WB (1972) Role of the Host Cell in Bacteriophage Morphogenesis: Effects of a Bacterial Mutation on T4 Head Assembly. Nature New Biol. 239:38–41

Georgopoulos CP, Herskowitz I (1971) *Escherichia coli* mutants blocked in lambda DNA synthesis. In: Hershey AD (ed) The bacteriophage lambda. Cold Spring Harbor Laboratory, Cold Spring Harbor, New York

Gething MJ (1999) Role and regulation of the ER chaperone BiP. Semin Cell Dev Biol 10:465–72

Gilinger G, Alwine JC (1993) Transcriptional activation by simian virus 40 large T antigen: requirements for simple promoter structures containing either TATA or initiator elements with variable upstream factor binding sites. J Virol 67:6682–6688

Glotzer JB, Saltik M, Chiocca S, Michou AI, Moseley P, Cotten M (2000) Activation of heat-shock response by an adenovirus is essential for virus replication. Nature 407:207–211

Goh PY, Tan YJ, Lim SP, Lim SG, Tan YH, Hong WJ (2001) The hepatitis C virus core protein interacts with NS5A and activates its caspase-mediated proteolytic cleavage. Virology 290:224–236

Gowda S, Satyanarayana T, Ayllon MA, Moreno P, Flores R, Dawson WO (2003) The conserved structures of the 5' nontranslated region of Citrus tristeza virus are involved in replication and virion assembly. Virology 317:50–64

Greene LE, Eisenberg E (1990) Dissociation of clathrin from coated vesicles by the uncoating ATPase. J. Biol. Chem. 265:6682–6687

Greene MK, Maskos K, Landry SJ (1998) Role of the J-domain in the cooperation of Hsp40 with Hsp70. Proc Natl Acad Sci USA 95:6108–6113

Guerrero CA, Bouyssounade D, Zarate S, Isa P, Lopez T, Espinosa R, Romero P, Mendez E, Lopez S, Arias CF (2002) Heat shock cognate protein 70 is involved in rotavirus cell entry. J Virol 76:4096–4102

Gurer C, Cimarelli A, Luban J (2002) Specific incorporation of heat shock protein 70 family members into primate lentiviral virions. J Virol 76:4666–4670

Gyoo Park S, Kyung Rho J, Jung G (2002) Hsp90 makes the human HBV Pol competent for in vitro priming rather than maintaining the human HBV Pol/pregenomic RNA complex. Arch Biochem Biophys 401:99–107

Hagemeier C, Walker S, Caswell R, Kouzarides T, Sinclair J (1992) The human cytomegalovirus 80-kilodalton but not the 72-kilodalton immediate-early protein transactivates heterologous promoters in a TATA box-dependent mechanism and interacts directly with TFIID. J Virol 66:4452–4456

Hammond C, Helenius A (1994) Folding of VSV G protein: sequential interaction with BiP and calnexin. Science 266:456–458

Hantschel M, Pfister K, Jordan A, Scholz R, Andreesen R, Schmitz G, Schmetzer H, Hiddemann W, Multhoff G (2000) Hsp70 plasma membrane expression on primary tumor biopsy material and bone marrow of leukemic patients. Cell Stress Chaperones 5:438–442

Hardwick JM (2001) Apoptosis in viral pathogenesis. Cell Death Differ 8:109–110

Harris KF, Christensen JB, Radany EH, Imperiale MJ (1998) Novel mechanisms of E2F induction by BK virus large-T antigen: requirement of both the pRb-binding and the J domains. Mol Cell Biol 18:1746–1756

Harrison CJ, Hayer-Hartl M, Di Liberto M, Hartl F-U, Kuriyan J (1997) Crystal structure of the nucleotide exchange factor GrpE bound to the ATPase domain of the molecular chaperone DnaK. Science 276:431–435

Hartl FU, Hayer-Hartl M (2002) Molecular chaperones in the cytosol: from nascent chain to folded protein. Science 295:1852–1858

Hatakeyama S, Yada M, Matsumoto M, Ishida N, Nakayama KI (2001) U box proteins as a new family of ubiquitin-protein ligases. J Biol Chem 276:33111–33120

Helmbrecht K, Rensing L (1999) Different constitutive heat shock protein 70 expression during proliferation and differentiation of rat C6 glioma cells. Neurochem Res 24:1293–1299

Henke A, Launhardt H, Klement K, Stelzner A, Zell R, Munder T (2000) Apoptosis in coxsackievirus B3-caused diseases: interaction between the capsid protein VP2 and the proapoptotic protein siva. J Virol 74:4284–4290

Henke A, Nestler M, Strunze S, Saluz HP, Hortschansky P, Menzel B, Martin U, Zell R, Stelzner A, Munder T (2001) The apoptotic capability of coxsackievirus B3 is influenced by the efficient interaction between the capsid protein VP2 and the proapoptotic host protein Siva. Virology 289:15–22

Herrmann CH, Dery CV, Mathews MB (1987) Transactivation of host and viral genes by the adenovirus E1B 19 K tumor antigen. Oncogene 2:25–35

Hoff KG, Ta DT, Tapley TL, Silberg JJ, Vickery LE (2002) Hsc66 substrate specificity is directed toward a discrete region of the iron-sulfur cluster template protein IscU. J Biol Chem 277:27353–27359

Hoffmann HJ, Lyman SK, Lu C, Petit MA, Echols H (1992) Activity of the Hsp70 chaperone complex–DnaK, DnaJ, and GrpE in initiating phage lambda DNA replication by sequestering and releasing lambda P protein. Proc Natl Acad Sci USA 89:12108–12111

Hogue BG, Nayak DP (1992) Synthesis and processing of the influenza virus neuraminidase, a type II transmembrane glycoprotein. Virology 188:510–517

Höhfeld J, Jentsch S (1997) GrpE-like regulation of the Hsc70 chaperone by the anti-apoptotic protein BAG-1. EMBO J 16:6209–6216

Höhfeld J, Minami Y, Hartl FU (1995) Hip, a novel cochaperone involved in the eukaryotic Hsc70/Hsp40 reaction cycle. Cell 83:589–598

Hoshino H, Shimoyama M, Miwa M, Sugimura T (1983) Detection of lymphocytes producing a human retrovirus associated with adult T-cell leukemia by syncytia induction assay. Proc Natl Acad Sci USA 80:7337–7341

Hu J, Anselmo D (2000) In vitro reconstitution of a functional duck hepatitis B virus reverse transcriptase: posttranslational activation by Hsp90. J Virol 74:11447–11455

Hu J, Seeger C (1996) Hsp90 is required for the activity of a hepatitis B virus reverse transcriptase. Proc Natl Acad Sci USA 93:1060–1064

Hu J, Toft D, Anselmo D, Wang X (2002) In vitro reconstitution of functional hepadnavirus reverse transcriptase with cellular chaperone proteins. J Virol 76:269–279

Hu J, Toft DO, Seeger C (1997) Hepadnavirus assembly and reverse transcription require a multi-component chaperone complex which is incorporated into nucleocapsids. EMBO J 16:59–68

Ishiai M, Wada C, Kawasaki Y, Yura T (1994) Replication initiator protein RepE of mini-F plasmid: functional differentiation between monomers (initiator) and dimers (autogenous repressor). Proc Natl Acad Sci USA 91:3839–3843

Jäättelä M, Wissing D, Kokholm K, Kallunki T, Egeblad M (1998) Hsp70 exerts its anti-apoptotic function downstream of caspase-3-like proteases. EMBO J 17:6124–6134

Jamora C, Dennert G, Lee AS (1996) Inhibition of tumor progression by suppression of stress protein GRP78/BiP induction in fibrosarcoma B/C10ME. Proc Natl Acad Sci USA 93:7690–7694

Jiang J, Ballinger CA, Wu Y, Dai Q, Cyr DM, Hohfeld J, Patterson C (2001) CHIP is a U-box-dependent E3 ubiquitin ligase: Identification of Hsc70 as a target for ubiquitylation. J Biol Chem 13:13

Jindal S, Malkovsky M (1994) Stress responses to viral infection. Trends Microbiol 2:89–91

Jindal S, Young RA (1992) Vaccinia virus infection induces a stress response that leads to association of Hsp70 with viral proteins. J Virol 66:5357–5362

Jockusch H, Wiegand C, Mersch B, Rajes D (2001) Mutants of tobacco mosaic virus with temperature-sensitive coat proteins induce heat shock response in tobacco leaves. Mol Plant Microbe Interact 14:914–917

Johnson AD, Tytell M (1993) Exogenous HSP70 becomes cell associated, but not internalized, by stressed arterial smooth muscle cells. In Vitro Cell Dev Biol Anim 29A:807–812

Jolly C, Morimoto RI (2000) Role of the heat shock response and molecular chaperones in oncogenesis and cell death. J Natl Cancer Inst 92:1564–1572

Jones KA (1997) Taking a new TAK on tat transactivation. Genes Dev 11:2593–2599

Kanelakis KC, Morishima Y, Dittmar KD, Galigniana MD, Takayama S, Reed JC, Pratt WB (1999) Differential effects of the hsp70-binding protein BAG-1 on glucocorticoid receptor folding by the hsp90-based chaperone machinery. J Biol Chem 274:34134–34140

Kanelakis KC, Murphy PJ, Galigniana MD, Morishima Y, Takayama S, Reed JC, Toft DO, Pratt WB (2000) hsp70 interacting protein Hip does not affect glucocorticoid receptor folding by the hsp90-based chaperone machinery except to oppose the effect of BAG-1. Biochemistry 39:14314–14321

Kao HT, Capasso O, Heintz N, Nevins JR (1985) Cell cycle control of the human HSP70 gene: implications for the role of a cellular E1A-like function. Mol Cell Biol 5:628–633

Kao HT, Nevins JR (1983) Transcriptional activation and subsequent control of the human heat shock gene during adenovirus infection. Mol Cell Biol 3:2058–2065

Karasev AV, Kashina AS, Gelfand VI, Dolja VV (1992) HSP70-related 65 kDa protein of beet yellows closterovirus is a microtubule-binding protein. FEBS Lett 304:12–14

Karzai AW, McMacken R (1996) A bipartite signaling mechanism involved in DnaJ-mediated activation of the *Escherichia coli* DnaK protein. J Biol Chem 271:11236–11246

Kaur J, Das SN, Srivastava A, Ralhan R (1998) Cell surface expression of 70 kDa heat shock protein in human oral dysplasia and squamous cell carcinoma: correlation with clinicopathological features. Oral Oncol 34:93–98

Kawasaki Y, Wada C, Yura T (1990) Roles of *Escherichia coli* heat shock proteins DnaK, DnaJ and GrpE in mini-F plasmid replication. Mol Gen Genet 220:277–282

Kelley WL (1998) The J-domain family and the recruitment of chaperone power. Trends Biochem Sci 23:222–227

Kelley WL (1999) Molecular chaperones: How J domains turn on Hsp70s. Curr Biol 9:R305–R308

Kelley WL, Georgopoulos C (1997) The T/t common exon of simian virus 40, JC, and BK polyomavirus T antigens can functionally replace the J-domain of the *Escherichia coli* DnaJ molecular chaperone. Proc Natl Acad Sci USA 94:3679–3684

Kelley WL, Landry SJ (1994) Chaperone power in a virus? Trends Biochem Sci 19:277–278

Khandjian EW, Turler H (1983) Simian virus 40 and polyoma virus induce synthesis of heat shock proteins in permissive cells. Mol Cell Biol 3:1–8

Kim SY, Sharma S, Hoskins JR, Wickner S (2002) Interaction of the DnaK and DnaJ chaperone system with a native substrate, P1 RepA. J Biol Chem 277:44778–44783

King FW, Wawrzynow A, Hohfeld J, Zylicz M (2001) Co-chaperones Bag-1, Hop and Hsp40 regulate Hsc70 and Hsp90 interactions with wild-type or mutant p53. EMBO J 20:6297–6305

Kingston RE, Cowie A, Morimoto RI, Gwinn KA (1986) Binding of polyomavirus large T antigen to the human hsp70 promoter is not required for trans activation. Mol Cell Biol 6:3180–3190

Kishi A, Ichinohe T, Hirai I, Kamiguchi K, Tamura Y, Kinebuchi M, Torigoe T, Ichimiya S, Kondo N, Ishitani K, Yoshikawa T, Kondo M, Matsuura A, Sato N (2001) The cell surface-expressed HSC70-like molecule preferentially reacts with the rat T-cell receptor Vdelta6 family. Immunogenetics 53:401–409

Kitay MK, Rowe DT (1996) Protein-protein interactions between Epstein-Barr virus nuclear antigen-LP and cellular gene products: binding of 70-kilodalton heat shock proteins. Virology 220:91–99

Kobayashi K, Ohgitani E, Tanaka Y, Kita M, Imanishi J (1994) Herpes simplex virus-induced expression of 70 kDa heat shock protein (HSP70) requires early protein synthesis but not viral DNA replication. Microbiol Immunol 38:321–325

Komori H, Matsunaga F, Higuchi Y, Ishiai M, Wada C, Miki K (1999) Crystal structure of a prokaryotic replication initiator protein bound to DNA at 2.6 Å resolution. EMBO J 18:4597–4607

Kozutsumi Y, Segal M, Normington K, Gething MJ, Sambrook J (1988) The presence of malfolded proteins in the endoplasmic reticulum signals the induction of glucose-regulated proteins. Nature 332:462–464

Kraus VB, Inostroza JA, Yeung K, Reinberg D, Nevins JR (1994) Interaction of the Dr1 inhibitory factor with the TATA binding protein is disrupted by adenovirus E1A. Proc Natl Acad Sci USA 91:6279–6282

Kroll J (2002) Molecular chaperones and the process of cellular immortalization in vitro. Biogerontology 3:183–185

Lambert C, Prange R (2003) Chaperone action in the posttranslational topological reorientation of the hepatitis B virus large envelope protein: Implications for translocational regulation. Proc Natl Acad Sci USA 100:5199–5204

LaThangue NB, Latchman DS (1987) Nuclear accumulation of a heat-shock 70-like protein during herpes simplex virus replication. Biosci Rep 7:475–483

LaThangue NB, Shriver K, Dawson C, Chan WL (1984) Herpes simplex virus infection causes the accumulation of a heat-shock protein. EMBO J 3:267–277

Laufen T, Mayer MP, Beisel C, Klostermeier D, Reinstein J, Bukau B (1999) Mechanism of regulation of Hsp70 chaperones by DnaJ co-chaperones. Proc Natl Acad Sci USA 96:5452–5457

Laufen T, Zuber U, Buchberger A, Bukau B (1998) DnaJ proteins. In: Fink AL, Goto Y (eds) Molecular chaperones in proteins: Structure, function, and mode of action. Marcel Dekker, New York, pp 241–274

Le Seyec J, Chouteau P, Cannie I, Guguen-Guillouzo C, Gripon P (1999) Infection process of the hepatitis B virus depends on the presence of a defined sequence in the pre-S1 domain. J Virol 73:2052–2057

LeBowitz JH, McMacken R (1984) Initiation of DNA synthesis on single-stranded DNA templates in vitro promoted by the bacteriophage lambda O and P replication proteins. Adv Exp Med Biol 179:77–89

Lee TG, Tang N, Thompson S, Miller J, Katze MG (1994) The 58,000-dalton cellular inhibitor of the interferon-induced double-stranded RNA-activated protein kinase (PKR) is a member of the tetratricopeptide repeat family of proteins. Mol Cell Biol 14:2331–2342

Lee TG, Tomita J, Hovanessian AG, Katze MG (1990) Purification and partial characterization of a cellular inhibitor of the interferon-induced protein kinase of Mr 68,000 from influenza virus-infected cells. Proc Natl Acad Sci USA 87:6208–6212

Lee TG, Tomita J, Hovanessian AG, Katze MG (1992) Characterization and regulation of the 58,000-dalton cellular inhibitor of the interferon-induced, dsRNA-activated protein kinase. J Biol Chem 267:14238–14243

Leone G, Coffey MC, Gilmore R, Duncan R, Maybaum L, Lee PW (1996) C-terminal trimerization, but not N-terminal trimerization, of the reovirus cell attachment protein Is a posttranslational and Hsp70/ATP-dependent process. J Biol Chem 271:8466–8471

Li C-Y, Lee J-S, Ko Y-G, Kim J-I, Seo J-S (2000) Heat shock protein 70 inhibits apoptosis downstream of cytochrome *c* release and upstream of caspase-3 activation. J Biol Chem 275:25665–25671

Li H, Soderbarg K, Houshmand H, You ZY, Magnusson G (2001) Effect on polyomavirus T-antigen function of mutations in a conserved leucine-rich segment of the DnaJ domain. J Virol 75:2253–2261

Liberek K, Georgopoulos C, Zylicz M (1988) Role of the *Escherichia coli* DnaK and DnaJ heat shock proteins in the initiation of bacteriophage lambda DNA replication. Proc Natl Acad Sci USA 85:6632–6636

Liberek K, Marszalek J, Ang D, Georgopoulos C, Zylicz M (1991) *Escherichia coli* DnaJ and GrpE heat shock proteins jointly stimulate ATPase activity of DnaK. Proc Natl Acad Sci USA 88:2874–2878

Liberek K, Osipiuk J, Zylicz M, Ang D, Skorko J, Georgopoulos C (1990) Physical interactions between bacteriophage and *Escherichia coli* proteins required for initiation of lambda DNA replication. J Biol Chem 265:3022–3029

Liberman E, Fong YL, Selby MJ, Choo QL, Cousens L, Houghton M, Yen TS (1999) Activation of the grp78 and grp94 promoters by hepatitis C virus E2 envelope protein. J Virol 73:3718–3722

Lin BY, Makhov AM, Griffith JD, Broker TR, Chow LT (2002) Chaperone proteins abrogate inhibition of the human papillomavirus (HPV) E1 replicative helicase by the HPV E2 protein. Mol Cell Biol 22:6592–6604

Lipsker D, Ziylan U, Spehner D, Proamer F, Bausinger H, Jeannin P, Salamero J, Bohbot A, Cazenave JP, Drillien R, Delneste Y, Hanau D, de la Salle H (2002) Heat shock proteins 70 and 60 share common receptors which are expressed on human monocyte-derived but not epidermal dendritic cells. Eur J Immunol 32:322–332

Liu JS, Kuo SR, Makhov AM, Cyr DM, Griffith JD, Broker TR, Chow LT (1998) Human Hsp70 and Hsp40 chaperone proteins facilitate human papillomavirus-11 E1 protein binding to the origin and stimulate cell-free DNA replication. J Biol Chem 273:30704–30710

Löffler-Mary H, Werr M, Prange R (1997) Sequence-specific repression of cotranslational translocation of the hepatitis B virus envelope proteins coincides with binding of heat shock protein Hsc70. Virology 235:144–52

Lowe SW, Ruley HE (1993) Stabilization of the p53 tumor suppressor is induced by adenovirus 5 E1A and accompanies apoptosis. Genes Dev 7:535–545

Lüders J, Demand J, Höhfeld J (2000) The ubiquitin-related BAG-1 provides a link between the molecular chaperones Hsc70/Hsp70 and the proteasome. J Biol Chem 275:4613–4617

Lüders J, Demand J, Schönfelder S, Frien M, Zimmermann R, Höhfeld J (1998) Cofactor-induced modulation of the functional specificity of the molecular chaperone Hsc70. Biol Chem 379:1217–1226

Lum LS, Hsu S, Vaewhongs M, Wu B (1992) The hsp70 gene CCAAT-binding factor mediates transcriptional activation by the adenovirus E1a protein. Mol Cell Biol 12:2599–2605

Macejak DG, Luftig RB (1991) Association of HSP70 with the adenovirus type 5 fiber protein in infected HEp-2 cells. Virology 180:120–125

Macejak DG, Sarnow P (1992) Association of heat shock protein 70 with enterovirus capsid precursor P1 in infected human cells. J Virol 66:1520–1527

Machamer CE, Doms RW, Bole DG, Helenius A, Rose JK (1990) Heavy chain binding protein recognizes incompletely disulfide-bonded forms of vesicular stomatitis virus G protein. J Biol Chem 265:6879–6883

Mallory JB, Alfano C, McMacken R (1990) Host virus interactions in the initiation of bacteriophage lambda DNA replication. Recruitment of *Escherichia coli* DnaB helicase by lambda P replication protein. J Biol Chem 265:13297–13307

Manara GC, Sansoni P, Badiali-De Giorgi L, Gallinella G, Ferrari C, Brianti V, Fagnoni FF, Ruegg CL, De Panfilis G, Pasquinelli G (1993) New insights suggesting a possible role of a heat shock protein 70-kD family-related protein in antigen processing/presentation phenomenon in humans. Blood 82:2865–2871

Mao H, Palmer D, Rosenthal KS (2001) Changes in BiP (GRP78) levels upon HSV-1 infection are strain dependent. Virus Res 76:127–135

Matsunaga F, Ishiai M, Kobayashi G, Uga H, Yura T, Wada C (1997) The central region of RepE initiator protein of mini-F plasmid plays a crucial role in dimerization required for negative replication control. J Mol Biol 274:27–38

Mayer MP, Laufen T, Paal K, McCarty JS, Bukau B (1999) Investigation of the interaction between DnaK and DnaJ by surface plasmon resonance spectroscopy. J Mol Biol 289:1131–1144

Mayer MP, Nikolay R, Bukau B (2002) Aha, another regulator for hsp90 chaperones. Mol Cell 10:1255–1256

Mayer MP, Rüdiger S, Bukau B (2000a) Molecular basis for interactions of the DnaK chaperone with substrates. Biol Chem 381:877–885

Mayer MP, Schröder H, Rüdiger S, Paal K, Laufen T, Bukau B (2000b) Multistep mechanism of substrate binding determines chaperone activity of Hsp70. Nat Struct Biol 7:586–593

Meacham GC, Patterson C, Zhang W, Younger JM, Cyr DM (2001) The Hsc70 co-chaperone CHIP targets immature CFTR for proteasomal degradation. Nat Cell Biol 3:100–105

Medina V, Peremyslov VV, Hagiwara Y, Dolja VV (1999) Subcellular localization of the HSP70-homolog encoded by beet yellows closterovirus. Virology 260:173–181

Melville MW, Hansen WJ, Freeman BC, Welch WJ, Katze MG (1997) The molecular chaperone hsp40 regulates the activity of $P58^{IPK}$, the cellular inhibitor of PKR. Proc Natl Acad Sci USA 94:97–102

Melville MW, Tan SL, Wambach M, Song J, Morimoto RI, Katze MG (1999) The cellular inhibitor of the PKR protein kinase, P58(IPK), is an influenza virus-activated co-chaperone that modulates heat shock protein 70 activity. J Biol Chem 274:3797–3803

Mensa-Wilmot K, Seaby R, Alfano C, Wold MS, Gomes B, McMacken R (1989) Reconstitution of a nine-protein system that initiates bacteriophage λ DNA replication. J Biol Chem 264:2853–2861

Milani V, Noessner E, Ghose S, Kuppner M, Ahrens B, Scharner A, Gastpar R, Issels RD (2002) Heat shock protein 70: role in antigen presentation and immune stimulation. Int J Hyperthermia 18:563–575

Milarski KL, Morimoto RI (1986) Expression of human HSP70 during the synthetic phase of the cell cycle. Proc Natl Acad Sci USA 83:9517–9521

Miller D, Brough S, al-Harbi O (1992) Characterization and cellular distribution of human spermatozoal heat shock proteins. Hum Reprod 7:637–645

Mirazimi A, Svensson L (2000) ATP is required for correct folding and disulfide bond formation of rotavirus VP7. J Virol 74:8048–8052

Misselwitz B, Staeck O, Rapoport TA (1998) J proteins catalytically activate Hsp70 molecules to trap a wide range of peptide sequences. Mol Cell 2:593–603

Moehler M, Zeidler M, Schede J, Rommelaere J, Galle PR, Cornelis JJ, Heike M (2003) Oncolytic parvovirus H1 induces release of heat-shock protein HSP72 in susceptible human tumor cells but may not affect primary immune cells. Cancer Gene Ther 10:477–480

Mokranjac D, Sichting M, Neupert W, Hell K (2003) Tim14, a novel key component of the import motor of the TIM23 protein translocase of mitochondria. EMBO J 22:4945–4956

Moore M, Schaack J, Baim SB, Morimoto RI, Shenk T (1987) Induced heat shock mRNAs escape the nucleocytoplasmic transport block in adenovirus-infected HeLa cells. Mol Cell Biol 7:4505–4512

Moratilla M, Agromayor M, Nunez A, Funes JM, Varas AJ, Lopez-Estebaranz JL, Esteban M, Martin-Gallardo A (1997) A random DNA sequencing, computer-based approach for the generation of a gene map of molluscum contagiosum virus. Virus Genes 14:73–80

Morimoto RI (1999) Inducible transcriptional regulation of heat shock genes: the stress signal and the unfolded protein response. In: Bukau B (ed) Molecular chaperones and folding catalysts. Regulation, cellular function and mechanism. Harwood Academic Publishers, Amsterdam, pp 35–52

Morimoto RI, Tissieres A, Georgopoulos C, eds. (1994) The biology of heat shock proteins and molecular chaperones. Cold Spring Harbor Laboratory Press, Cold Spring Harbor, NY

Mosser DD, Caron AW, Bourget L, Denis-Larose C, Massie B (1997) Role of the human heat shock protein hsp70 in protection against stress-induced apoptosis. Mol Cell Biol 17:5317–5327

Mosser DD, Caron AW, Bourget L, Meriin AB, Sherman MY, Morimoto RI, Massie B (2000) The chaperone function of hsp70 is required for protection against stress-induced apoptosis. Mol Cell Biol 20:7146–7159

Multhoff G, Botzler C, Issels R (1998) The Role of Heat Shock Proteins in the Stimulation of an Immune Response. Biol Chem 379:295–300

Multhoff G, Hightower LE (1996) Cell surface expression of heat shock proteins and the immune response. Cell Stress Chaperones 1:167–176

Mulvey M, Brown DT (1995) Involvement of the molecular chaperone BiP in maturation of Sindbis virus envelope glycoproteins. J Virol 69:1621–1627

Muthumani K, Hwang DS, Desai BM, Zhang D, Dayes N, Green DR, Weiner DB (2002a) HIV-1 Vpr induces apoptosis through caspase 9 in T cells and peripheral blood mononuclear cells. J Biol Chem 277:37820–37831

Muthumani K, Zhang D, Hwang DS, Kudchodkar S, Dayes NS, Desai BM, Malik AS, Yang JS, Chattergoon MA, Maguire HC, Jr., Weiner DB (2002b) Adenovirus encoding HIV-1 Vpr activates caspase 9 and induces apoptotic cell death in both p53 positive and negative human tumor cell lines. Oncogene 21:4613–4625

Napuli AJ, Alzhanova DV, Doneanu CE, Barofsky DF, Koonin EV, Dolja VV (2003) The 64-kilodalton capsid protein homolog of Beet yellows virus is required for assembly of virion tails. J Virol 77:2377–2384

Napuli AJ, Falk BW, Dolja VV (2000) Interaction between HSP70 homolog and filamentous virions of the Beet yellows virus. Virology 274:232–239

Neupert W, Brunner M (2002) The protein import motor of mitochondria. Nat Rev Mol Cell Biol 3:555–565

Nevins JR (1982) Induction of the synthesis of a 70,000 dalton mammalian heat shock protein by the adenovirus E1A gene product. Cell 29:913–919

Newmyer SL, Christensen A, Sever S (2003) Auxilin-dynamin interactions link the uncoating ATPase chaperone machinery with vesicle formation. Dev Cell 4:929–940

Newmyer SL, Schmid SL (2001) Dominant-interfering Hsc70 mutants disrupt multiple stages of the clathrin-coated vesicle cycle in vivo. J Cell Biol 152:607–620

Ng DT, Randall RE, Lamb RA (1989) Intracellular maturation and transport of the SV5 type II glycoprotein hemagglutinin-neuraminidase: specific and transient association with GRP78-BiP in the endoplasmic reticulum and extensive internalization from the cell surface. J Cell Biol 109:3273–3289

Nicchitta CV (2000) Role of chaperones in antigen processing. Immunol Invest 29:101–104

Niewiarowska J, D'Halluin JC, Belin MT (1992) Adenovirus capsid proteins interact with HSP70 proteins after penetration in human or rodent cells. Exp Cell Res 201:408–416

Niyaz Y, Frenz I, Petersen G, Gehring U (2003) Transcriptional stimulation by the DNA binding protein Hap46/BAG-1M involves hsp70/hsc70 molecular chaperones. Nucl Acids Res 31:2209–2216

Niyaz Y, Zeiner M, Gehring U (2001) Transcriptional activation by the human Hsp70-associating protein Hap50. J Cell Sci 114:1839–1845

Nollen EAA, Brunsting JF, Song J, Kampinga HH, Morimoto RI (2000) Bag1 functions *in vivo* as a negative regulator of Hsp70 chaperone activity. Mol Cell Biol 20:1083–1088

Nylandsted J, Rohde M, Brand K, Bastholm L, Elling F, Jaattela M (2000) Selective depletion of heat shock protein 70 (Hsp70) activates a tumor-specific death program that is independent of caspases and bypasses Bcl-2. Proc Natl Acad Sci USA 97:7871–7876

O'Keeffe B, Fong Y, Chen D, Zhou S, Zhou Q (2000) Requirement for a kinase-specific chaperone pathway in the production of a Cdk9/cyclin T1 heterodimer responsible for P-TEFb-mediated tat stimulation of HIV-1 transcription. J Biol Chem 275:279–287

Oglesbee M, Krakowka S (1993) Cellular stress response induces selective intranuclear trafficking and accumulation of morbillivirus major core protein. Lab Invest 68:109–117

Oglesbee M, Ringler S, Krakowka S (1990) Interaction of canine distemper virus nucleocapsid variants with 70 K heat-shock proteins. J Gen Virol 71:1585–1590

Oglesbee MJ, Liu Z, Kenney H, Brooks CL (1996) The highly inducible member of the 70 kDa family of heat shock proteins increases canine distemper virus polymerase activity. J Gen Virol 77:2125–2135

Ohgitani E, Kobayashi K, Takeshita K, Imanishi J (1998) Induced expression and localization to nuclear-inclusion bodies of hsp70 in varicella-zoster virus-infected human diploid fibroblasts. Microbiol Immunol 42:755–760

Ohgitani E, Kobayashi K, Takeshita K, Imanishi J (1999) Biphasic translocation of a 70 kDa heat shock protein in human cytomegalovirus-infected cells. J Gen Virol 80:63–68

Otteken A, Earl PL, Moss B (1996) Folding, assembly, and intracellular trafficking of the human immunodeficiency virus type 1 envelope glycoprotein analyzed with monoclonal antibodies recognizing maturational intermediates. J Virol 70:3407–3415

Packschies L, Theyssen H, Buchberger A, Bukau B, Goody RS, Reinstein J (1997) GrpE accelerates nucleotide exchange of the molecular chaperone DnaK with an associative displacement mechanism. Biochemistry 36:3417–3422

Palleros DR, Reid KL, Shi L, Welch WJ, Fink AL (1993) ATP-induced protein-Hsp70 complex dissociation requires K^+ but not ATP hydrolysis. Nature 365:664–666

Panjwani N, Akbari O, Garcia S, Brazil M, Stockinger B (1999) The HSC73 molecular chaperone: involvement in MHC class II antigen presentation. J Immunol 163:1936–1942

Parks GD, Lamb RA (1990) Defective assembly and intracellular transport of mutant paramyxovirus hemagglutinin-neuraminidase proteins containing altered cytoplasmic domains. J Virol 64:3605–3616

Patrick DR, Oliff A, Heimbrook DC (1994) Identification of a novel retinoblastoma gene product binding site on human papillomavirus type 16 E7 protein. J Biol Chem 269:6842–6850

Paulsson K, Wang P (2003) Chaperones and folding of MHC class I molecules in the endoplasmic reticulum. Biochim Biophys Acta 1641:1–12

Pearl LH, Prodromou C (2002) Structure, function, and mechanism of the Hsp90 molecular chaperone. Adv Protein Chem 59:157–186

Peluso RW, Lamb RA, Choppin PW (1978) Infection with paramyxoviruses stimulates synthesis of cellular polypeptides that are also stimulated in cells transformed by Rous sarcoma virus or deprived of glucose. Proc Natl Acad Sci USA 75:6120–6124

Peremyslov VV, Dolja VV (2002) Identification of the subgenomic mRNAs that encode 6-kDa movement protein and Hsp70 homolog of Beet yellows virus. Virology 295:299–306

Peremyslov VV, Hagiwara Y, Dolja VV (1999) HSP70 homolog functions in cell-to-cell movement of a plant virus. Proc Natl Acad Sci USA 96:14771–14776

Phillips B, Abravaya K, Morimoto RI (1991) Analysis of the specificity and mechanism of transcriptional activation of the human hsp70 gene during infection by DNA viruses. J Virol 65:5680–5692

Phillips B, Morimoto RI (1991) Transcriptional regulation of human hsp70 genes: relationship between cell growth, differentiation, virus infection, and the stress response. Results Probl Cell Differ 17:167–187

Pierce SK (1994) Molecular chaperones in the processing and presentation of antigen to helper T cells. Experientia 50:1026–1030

Pierpaoli EV, Sandmeier E, Baici A, Schönfeld H-J, Gisler S, Christen P (1997) The power stroke of the DnaK/DnaJ/GrpE molecular chaperone system. J Mol Biol 269:757–768

Pollack JR, Ganem D (1994) Site-specific RNA binding by a hepatitis B virus reverse transcriptase initiates two distinct reactions: RNA packaging and DNA synthesis. J Virol 68:5579–5587

Prange R, Werr M, Löffler-Mary H (1999) Chaperones involved in hepatitis B virus morphogenesis. Biol Chem 380:305–314

Prapapanich V, Chen S, Nair S, Rimerman R, Smith D (1996) Molecular cloning of human p48, a transient component of progesterone receptor complexes and an Hsp70-binding protein. Mol Endocrinol. 10:420–431

Pratt WB (1997) The role of the hsp90-based chaperone system in signal transduction by nuclear receptors and receptors signaling via MAP kinase. Annu Rev Pharmacol Toxicol 37:297–326

Pratt WB, Toft DO (2003) Regulation of signaling protein function and trafficking by the hsp90/hsp70-based chaperone machinery. Exp Biol Med (Maywood) 228:111–33

Prokhnevsky AI, Peremyslov VV, Napuli AJ, Dolja VV (2002) Interaction between long-distance transport factor and Hsp70-related movement protein of Beet yellows virus. J Virol 76:11003–11011

Qian YQ, Patel D, Hartl FU, McColl DJ (1996) Nuclear magnetic resonance solution structure of the human Hsp40 (HDJ-1) J-domain. J Mol Biol 260:224–235

Queitsch C, Sangster TA, Lindquist S (2002) Hsp90 as a capacitor of phenotypic variation. Nature 417:618–624

Rapoport TA, Jungnickel B, Kutay U (1996) Protein transport across the eukaryotic endoplasmic reticulum and bacterial inner membranes. Annu Rev Biochem 65:271–303

Ravagnan L, Gurbuxani S, Susin SA, Maisse C, Daugas E, Zamzami N, Mak T, Jaattela M, Penninger JM, Garrido C, Kroemer G (2001) Heat-shock protein 70 antagonizes apoptosis-inducing factor. Nat Cell Biol 3:839–843

Richter K, Buchner J (2001) Hsp90: chaperoning signal transduction. J Cell Physiol 188:281–290

Rivas C, Gil J, Melkova Z, Esteban M, Diaz-Guerra M (1998) Vaccinia virus E3L protein is an inhibitor of the interferon (IFN)-induced 2–5A synthetase enzyme. Virology 243:406–414

Roberts JD, McMacken R (1983) The bacteriophage lambda O replication protein: isolation and characterization of the amplified initiator. Nucl Acids Res 11:7435–7452

Rocchi G, Pavesi A, Ferrari C, Bolchi A, Manara GC (1993) A new insight into the suggestion of a possible antigenic role of a member of the 70 kD heat shock proteins. Cell Biol Int 17:83–92

Roth S, Willcox N, Rzepka R, Mayer MP, Melchers I (2002) Major differences in antigen-processing correlate with a single Arg71<–>Lys substitution in HLA-DR molecules predisposing to rheumatoid arthritis and with their selective interactions with 70-kDa heat shock protein chaperones. J Immunol 169:3015–3020

Roulston A, Marcellus RC, Branton PE (1999) Viruses and apoptosis. Annu Rev Microbiol 53:577–628

Roux L (1990) Selective and transient association of Sendai virus HN glycoprotein with BiP. Virology 175:161–166

Rüdiger S, Germeroth L, Schneider-Mergener J, Bukau B (1997) Substrate specificity of the DnaK chaperone determined by screening cellulose-bound peptide libraries. EMBO J 16:1501–1507

Rutherford SL, Lindquist S (1998) Hsp90 as a capacitor for morphological evolution. Nature 396:336–342

Ryan MT, Pfanner N (2002) Hsp70 proteins in protein translocation. Adv Protein Chem 59:223–242

Sagara J, Kawai A (1992) Identification of heat shock protein 70 in the rabies virion. Virology 190:845–848

Sagara Y, Ishida C, Inoue Y, Shiraki H, Maeda Y (1998) 71-kilodalton heat shock cognate protein acts as a cellular receptor for syncytium formation induced by human T-cell lymphotropic virus type 1. J Virol 72:535–541

Sainis I, Angelidis C, Pagoulatos G, Lazaridis I (1994) The hsc70 gene which is slightly induced by heat is the main virus inducible member of the hsp70 gene family. FEBS Lett 355:282–286

Saito H, Uchida H (1977) Initiation of the DNA replication of bacteriophage lambda in *Escherichia coli* K12. J Mol Biol 113:1–25

Saleh A, Srinivasula SM, Balkir L, Robbins PD, Alnemri ES (2000) Negative regulation of the Apaf-1 apoptosome by Hsp70. Nature Cell Biol 2:476–483

Santomenna LD, Colberg-Poley AM (1990) Induction of cellular hsp70 expression by human cytomegalovirus. J Virol 64:2033–2040

Saphire AC, Guan T, Schirmer EC, Nemerow GR, Gerace L (2000) Nuclear import of adenovirus DNA in vitro involves the nuclear protein import pathway and hsc70. J Biol Chem 275:4298–4304

Satyanarayana T, Gowda S, Ayllon MA, Dawson WO (2004) Closterovirus bipolar virion: evidence for initiation of assembly by minor coat protein and its restriction to the genomic RNA 5' region. Proc Natl Acad Sci USA 101:799–804

Satyanarayana T, Gowda S, Mawassi M, Albiach-Marti MR, Ayllon MA, Robertson C, Garnsey SM, Dawson WO (2000) Closterovirus encoded HSP70 homolog and p61 in addition to both coat proteins function in efficient virion assembly. Virology 278:253–265

Scheufler C, Brinker A, Bourenkov G, Pegoraro S, Moroder L, Bartunik H, Hartl FU, Moarefi I (2000) Struture of TPR domain-peptide complexes: critical elements in the assembly of the Hsp70-Hsp90 multichaperone machine. Cell 101:199–210

Schilling B, De-Medina T, Syken J, Vidal M, Münger K (1998) A novel human DnaJ protein, hTid-1, a homolog of the Drosophila tumor suppressor protein Tid56, can interact with the human papillomavirus type 16 E7 oncoprotein. Virology 247:74–85

Schirmbeck R, Kwissa M, Fissolo N, Elkholy S, Riedl P, Reimann J (2002) Priming polyvalent immunity by DNA vaccines expressing chimeric antigens with a stress protein-capturing, viral J-domain. FASEB J 16:1108–1110

Schlieker C, Bukau B, Mogk A (2002) Prevention and reversion of protein aggregation by molecular chaperones in the *E. coli* cytosol: implications for their applicability in biotechnology. J Biotechnol 96:13–21

Schmid D, Baici A, Gehring H, Christen P (1994) Kinetics of molecular chaperone action. Science 263:971–973

Schneikert J, Hubner S, Martin E, Cato AC (1999) A nuclear action of the eukaryotic cochaperone RAP46 in downregulation of glucocorticoid receptor activity. J Cell Biol 146:929–940

Schultz-Cherry S, Dybdahl-Sissoko N, Neumann G, Kawaoka Y, Hinshaw VS (2001) Influenza virus ns1 protein induces apoptosis in cultured cells. J Virol 75:7875–7881

Sedger L, Ramshaw I, Condie A, Medveczky J, Braithwaite A, Ruby J (1996) Vaccinia virus replication is independent of cellular HSP72 expression which is induced during virus infection. Virology 225:423–427

Sedger L, Ruby J (1994) Heat shock response to vaccinia virus infection. J Virol 68:4685–4689

Sharma S, Sathyanarayana BK, Bird JG, Hoskins JR, Lee B, Wickner S (2004) Plasmid P1 RepA is homologous to the F plasmid RepE class of initiators. J Biol Chem 279:6027–6034

Sheng Q, Denis D, Ratnofsky M, DeCaprio JA, Schaffhausen B (1997) The DnaJ domain of polyomavirus large T antigen is required to regulate RB family tumor suppressor function. J Virol 71:9410–9416

Sheng Q, Love TM, Schaffhausen B (2000) J domain-independent regulation of the Rb family by polyomavirus large T antigen. J Virol 74:5280–5290

Shin BK, Wang H, Yim AM, Le Naour F, Brichory F, Jang JH, Zhao R, Puravs E, Tra J, Michael CW, Misek DE, Hanash SM (2003) Global profiling of the cell surface proteome of cancer cells uncovers an abundance of proteins with chaperone function. J Biol Chem 278:7607–7616

Shtilerman M, Lorimer GH, Englander SW (1999) Chaperonin function: folding by forced unfolding. Science 284:822–825

Simon MC, Fisch TM, Benecke BJ, Nevins JR, Heintz N (1988) Definition of Multiple, Functionally Distinct TATA Elements, one of which is a target in the *hsp70* promoter for E1A Regulation. Cell 52:723–729

Simon MC, Kitchener K, Kao HT, Hickey E, Weber L, Voellmy R, Heintz N, Nevins JR (1987) Selective induction of human heat shock gene transcription by the adenovirus E1A gene products, including the 12S E1A product. Mol Cell Biol 7:2884–2890

Singh I, Doms RW, Wagner KR, Helenius A (1990) Intracellular transport of soluble and membrane-bound glycoproteins: folding, assembly and secretion of anchor-free influenza hemagglutinin. EMBO J 9:631–639

Slepenkov SV, Witt SN (2002) The unfolding story of the *Escherichia coli* Hsp70 DnaK: is DnaK a holdase or an unfoldase? Mol Microbiol 45:1197–1206

Smith DF (2000) Chaperones in progesterone receptor complexes. Semin. Cell Dev Biol 11:45–52

Smith DF, Sullivan WP, Marion TN, Zaitsu K, Madden B, McCormick DJ, Toft DO (1993) Identification of a 60-Kilodalton Stress-Related Protein, p60, which interacts with hsp90 and hsp70. Mol Cell Biol 13:869–876

Soellick T, Uhrig JF, Bucher GL, Kellmann JW, Schreier PH (2000) The movement protein NSm of tomato spotted wilt tospovirus (TSWV): RNA binding, interaction with the TSWV N protein, and identification of interacting plant proteins. Proc Natl Acad Sci USA 97:2373–2378

Sollars V, Lu X, Xiao L, Wang X, Garfinkel MD, Ruden DM (2003) Evidence for an epigenetic mechanism by which Hsp90 acts as a capacitor for morphological evolution. Nat Genet 33:70–74

Sondermann H, Becker T, Mayhew M, Wieland F, Hartl FU (2000) Characterization of a receptor for heat shock protein 70 on macrophages and monocytes. Biol Chem 381:1165–1174

Sondermann H, Scheufler C, Scheider C, Höhfeld J, Hartl F-U, Moarefi I (2001) Structure of a Bag/Hsc70 Complex: Convergent functional evolution of Hsp70 nucleotide exchange factors. Science 291:1553–1557

Sousa MC, McKay DB (1998) The hydroxyl of threonine 13 of the bovine 70-kDa heat shock cognate protein is essential for transducing the ATP-induced conformational change. Biochemistry 37:15392–15399

Srinivasan A, McClellan AJ, Vartikar J, Marks I, Cantalupo P, Li Y, Whyte P, Rundell K, Brodsky JL, Pipas JM (1997) The amino-terminal transforming region of simian virus 40 large T and small t antigens functions as a J domain. Mol Cell Biol 17:4761–4773

Stoeckle MY, Sugano S, Hampe A, Vashistha A, Pellman D, Hanafusa H (1988) 78-kilodalton glucose-regulated protein is induced in Rous sarcoma virus-transformed cells independently of glucose deprivation. Mol Cell Biol 8:2675–2680

Stubdal H, Zalvide J, Campbell KS, Schweitzer C, Roberts TM, DeCaprio JA (1997) Inactivation of pRB-related proteins p130 and p107 mediated by the J domain of simian virus 40 large T antigen. Mol Cell Biol 17:4979–4990

Sugawara S, Takeda K, Lee A, Dennert G (1993) Suppression of stress protein GRP78 induction in tumor B/C10ME eliminates resistance to cell mediated cytotoxicity. Cancer Res 53:6001–6005

Sullivan CS, Cantalupo P, Pipas JM (2000a) The molecular chaperone activity of simian virus 40 large T antigen is required to disrupt Rb-E2F family complexes by an ATP-dependent mechanism. Mol Cell Biol 20:6233–6243

Sullivan CS, Gilbert SP, Pipas JM (2001) ATP-dependent simian virus 40 T-antigen-Hsc70 complex formation. J Virol 75:1601–1610

Sullivan CS, Pipas JM (2002) T antigens of simian virus 40: molecular chaperones for viral replication and tumorigenesis. Microbiol Mol Biol Rev 66:179–202

Sullivan CS, Tremblay JD, Fewell SW, Lewis JA, Brodsky JL, Pipas JM (2000b) Species-specific elements in the large T-antigen J domain are required for cellular transformation and DNA replication by simian virus 40. Mol Cell Biol 20:5749–5757

Sunshine M, Feiss M, Stuart J, Yochem Y (1977) A new host gene (groPC) is necessary for lambda DNA replication. Mol Gen Genet 151:27–34

Suzue K, Zhou X, Eisen HN, Young RA (1997) Heat shock fusion proteins as vehicles for antigen delivery into the major histocompatibility complex class I presentation pathway. Proc Natl Acad Sci USA 94:13146–13151

Swack JA, Pal SK, Mason RJ, Abeles AL, Chattoraj DK (1987) P1 plasmid replication: measurement of initiator protein concentration in vivo. J Bacteriol 169:3737–3742

Symonds H, Krall L, Remington L, Saenz-Robles M, Lowe S, Jacks T, Van Dyke T (1994) p53-dependent apoptosis suppresses tumor growth and progression in vivo. Cell 78:703–711

Takahashi N, Sasaki R, Takahashi J, Takayama S, Reed JC, Andoh T (2001) BAG-1 M, an isoform of Bcl-2-interacting protein BAG-1, enhances gene expression driven by CMV promoter. Biochem Biophys Res Commun 286:807–814

Takayama S, Bimston DN, Matsuzawa S-i, Freeman BC, Aime-Sempe C, Xie Z, Morimoto RI, Reed JC (1997) BAG-1 modulates the chaperone activity of Hsp70/Hsc70. EMBO J 16:4887–4896

Takayama S, Reed JC (2001) Molecular chaperone targeting and regulation by BAG family proteins. Nat Cell Biol 3:E237–E241

Takayama S, Sato T, Krajewski S, Kochel K, Irie S, Millan JA, Reed JC (1995) Cloning and Functional Analysis of BAG-1: A Novel Bcl-2-Binding Protein with Anti-Cell Death Activity. Cell 80:279–284

Takayama S, Xie Z, Reed JC (1999) An evolutionarily conserved family of Hsp70/Hsc70 molecular chaperone regulators. J Biol Chem 274:781–786

Takeda S, McKay DB (1996) Kinetics of peptide binding to the bovine 70 kDa heat shock cognate protein, a molecular chaperone. Biochemistry 35:4636–4644

Tang NM, Ho CY, Katze MG (1996) The 58-kDa cellular inhibitor of the double stranded RNA-dependent protein kinase requires the tetratricopeptide repeat 6 and DnaJ motifs to stimulate protein synthesis in vivo. J Biol Chem 271:28660–28666

Tanguy Le Gac N, Boehmer PE (2002) Activation of the herpes simplex virus type-1 origin-binding protein (UL9) by heat shock proteins. J Biol Chem 277:5660–5666

Tavis JE, Perri S, Ganem D (1994) Hepadnavirus reverse transcription initiates within the stem-loop of the RNA packaging signal and employs a novel strand transfer. J Virol 68:3536–3543

Theodorakis NG, Morimoto RI (1987) Posttranscriptional regulation of hsp70 expression in human cells: effects of heat shock, inhibition of protein synthesis, and adenovirus infection on translation and mRNA stability. Mol Cell Biol 7:4357–4368

Theyssen H, Schuster H-P, Bukau B, Reinstein J (1996) The second step of ATP binding to DnaK induces peptide release. J Mol Biol 263:657–670

Tilly K, Spence J, Georgopoulos C (1989) Modulation of stability of the *Escherichia coli* heat shock regulatory factor σ^{32}. J Bacteriol 171:1585–1589

Tilly K, Yarmolinsky M (1989) Participation of *Escherichia coli* heat shock proteins DnaJ, DnaK, and GrpE in P1 plasmid replication. J Bacteriol 171:6025–6029

Tomita Y, Yamashita T, Sato H, Taira H (1999) Kinetics of interactions of sendai virus envelope glycoproteins, F and HN, with endoplasmic reticulum-resident molecular chaperones, BiP, calnexin, and calreticulin. J Biochem (Tokyo) 126:1090–1100

Triantafilou K, Fradelizi D, Wilson K, Triantafilou M (2002) GRP78, a coreceptor for coxsackievirus A9, interacts with major histocompatibility complex class I molecules which mediate virus internalization. J Virol 76:633–643

Truscott KN, Voos W, Frazier AE, Lind M, Li Y, Geissler A, Dudek J, Müller H, Sickmann A, Meyer HE, Meisinger C, Guiard B, Rehling P, Pfanner N (2003) A J-protein is an essential subunit of the presequence translocase-associated protein import motor of mitochondria. J Cell Biol 163:707–713

Ungewickell E (1985) The 70-kd mammalian heat shock proteins are structurally and functionally related to the uncoating protein that releases clathrin triskelia from coated vesicles. EMBO J 4:3385–3391

Urano F, Bertolotti A, Ron D (2000) IRE1 and efferent signaling from the endoplasmic reticulum. J Cell Sci 113:3697–3702

Vabulas RM, Ahmad-Nejad P, Ghose S, Kirschning CJ, Issels RD, Wagner H (2002) HSP70 as endogenous stimulus of the Toll/interleukin-1 receptor signal pathway. J Biol Chem 277:15107–15112

VanBuskirk A, Crump BL, Margoliash E, Pierce SK (1989) A peptide binding protein having a role in antigen presentation is a member of the HSP70 heat shock family. J Exp Med 170:1799–1809

VanBuskirk AM, DeNagel DC, Guagliardi LE, Brodsky FM, Pierce SK (1991) Cellular and subcellular distribution of PBP72/74, a peptide-binding protein that plays a role in antigen processing. J Immunol 146:500–556

Velten M, Villoutreix BO, Ladjimi MM (2000) Quaternary structure of HSC70 cochaperone HIP. Biochemistry 39:307–315

Wainberg Z, Oliveira M, Lerner S, Tao Y, Brenner BG (1997) Modulation of stress protein (hsp27 and hsp70) expression in CD4+ lymphocytic cells following acute infection with human immunodeficiency virus type-1. Virology 233:364–373

Wang GH, Seeger C (1992) The reverse transcriptase of hepatitis B virus acts as a protein primer for viral DNA synthesis. Cell 71:663–670

Wang GH, Seeger C (1993) Novel mechanism for reverse transcription in hepatitis B viruses. J Virol 67:6507–6512

Wang H-G, Takayam S, Rapp UR, Reed JC (1996) Bcl-2 interacting protein, BAG-1, binds to and activates the kinase Raf-1. Proc Natl Acad Sci USA 93:7063–7068

Watowich SS, Morimoto RI, Lamb RA (1991) Flux of the paramyxovirus hemagglutinin-neuraminidase glycoprotein through the endoplasmic reticulum activates transcription of the GRP78-BiP gene. J Virol 65:3590–3597

Weber-Ban EU, Reid BG, Miranker AD, Horwich AL (1999) Global unfolding of a substrate protein by the Hsp100 chaperone ClpA. Nature 401:90–93

Weisshart K, Bradley MK, Weiner BM, Schneider C, Moarefi I, Fanning E, Arthur AK (1996) An N-terminal deletion mutant of simian virus 40 (SV40) large T antigen oligomerizes incorrectly on SV40 DNA but retains the ability to bind to DNA polymerase alpha and replicate SV40 DNA in vitro. J Virol 70:3509–3516

White E, Spector D, Welch W (1988) Differential distribution of the adenovirus E1A proteins and colocalization of E1A with the 70-kilodalton cellular heat shock protein in infected cells. J Virol 62:4153–4166

Wickner S, Hoskins J, McKenney K (1991a) Function of DnaJ and DnaK as chaperones in origin-specific DNA binding by RepA. Nature 350:165–167

Wickner S, Hoskins J, McKenney K (1991b) Monomerization of RepA dimers by heat shock proteins activates binding to DNA replication origin. Proc Natl Acad Sci USA 88:7903–7907

Wickner S, Skowyra D, Hoskins J, McKenney K (1992) DnaJ, DnaK, and GrpE heat shock proteins are required in oriP1 DNA replication solely at the RepA monomerization step. Proc Natl Acad Sci USA 89:10345–10349

Wickner SH (1990) Three *Escherichia coli* heat shock proteins are required for P1 plasmid DNA replication: formation of an active complex between *E. coli* DnaJ protein and the P1 initiator protein. Proc Natl Acad Sci USA 87:2690–2694

Williams GT, McClanahan TK, Morimoto RI (1989) E1a transactivation of the human HSP70 promoter is mediated through the basal transcriptional complex. Mol Cell Biol 9:2574–2587

Wittung-Stafshede P, Guidry J, Horne BE, Landry SJ (2003) The J-domain of Hsp40 couples ATP hydrolysis to substrate capture in Hsp70. Biochemistry 42:4937–4944

Wu BJ, Hurst HC, Jones NC, Morimoto RI (1986) The E1A 13S product of adenovirus 5 activates transcription of the cellular human HSP70 gene. Mol Cell Biol 6:2994–2999

Wu EW, Clemens KE, Heck DV, Munger K (1993) The human papillomavirus E7 oncoprotein and the cellular transcription factor E2F bind to separate sites on the retinoblastoma tumor suppressor protein. J Virol 67:2402–2407

Wyman C, Vasilikiotis C, Ang D, Georgopoulos C, Echols H (1993) Function of the GrpE heat shock protein in bidirectional unwinding and replication from the origin of phage lambda. J Biol Chem 268:25192–25196

Xu A, Bellamy AR, Taylor JA (1998) BiP (GRP78) and endoplasmin (GRP94) are induced following rotavirus infection and bind transiently to an endoplasmic reticulum-localized virion component. J Virol 72:9865–9872

Yaglom JA, Gabai VL, Meriin AB, Mosser DD, Sherman MY (1999) The function of HSP72 in suppression of c-Jun N-terminal kinase activation can be dissociated from its role in prevention of protein damage. J Biol Chem 274:20223–20228

Yan W, Gale MJ, Jr, Tan SL, Katze MG (2002) Inactivation of the PKR protein kinase and stimulation of mRNA translation by the cellular co-chaperone P58(IPK) does not require J domain function. Biochemistry 41:4938–4945

Yang J, DeFranco DB (1994) Differential roles of heat shock protein 70 in the in vitro nuclear import of glucocorticoid receptor and simian virus 40 large tumor antigen. Mol Cell Biol 14:5088–5098

Yang UC, Huang W, Flint SJ (1996) mRNA export correlates with activation of transcription in human subgroup C adenovirus-infected cells. J Virol 70:4071–4080

Yochem J, Uchida H, Sunshine M, Saito H, Georgopoulos CP, Feiss M (1978) Genetic analysis of two genes, *dnaJ* and *dnaK*, necessary for *Escherichia coli* and bacteriophage lambda DNA replication. Mol Gen Genet 164:9–14

Young JC, Moarefi I, Hartl FU (2001) Hsp90: a specialized but essential protein-folding tool. J Cell Biol 154:267–273

Zalvide J, Stubdal H, DeCaprio JA (1998) The J domain of simian virus 40 large T antigen is required to functionally inactivate RB family proteins. Mol Cell Biol 18:1408–1415

Zarate S, Cuadras MA, Espinosa R, Romero P, Juarez KO, Camacho-Nuez M, Arias CF, Lopez S (2003) Interaction of rotaviruses with Hsc70 during cell entry is mediated by VP5. J Virol 77:7254–7260

Zeiner M, Gebauer M, Gehring U (1997) Mammalian protein RAP46: an interaction partner and modulator of 70 kDa heat shock proteins. EMBO J. 16:5483–5490

Zeiner M, Gehring U (1995) A protein that interacts with members of the nuclear hormone receptor family: Identification and cDNA cloning. Proc Natl Acad Sci USA 92:11465–11469

Zeiner M, Niyaz Y, Gehring U (1999) The hsp70-associating protein Hap46 binds to DNA and stimulates transcription. Proc Natl Acad Sci USA 96:10194–10199

Zhang X, Glendening C, Linke H, Parks CL, Brooks C, Udem SA, Oglesbee M (2002) Identification and characterization of a regulatory domain on the carboxyl terminus of the measles virus nucleocapsid protein. J Virol 76:8737–8746

Zhu X, Zhao X, Burkholder WF, Gragerov A, Ogata CM, Gottesman M, Hendrickson WA (1996) Structural analysis of substrate binding by the molecular chaperone DnaK. Science 272:1606–1614

Zylicz M, Ang D, Liberek K, Georgopoulos C (1989) Initiation of lambda DNA replication with purified host- and bacteriophage-encoded proteins: the role of the dnaK, dnaJ and grpE heat shock proteins. EMBO J 8:1601–1608

Zylicz M, Ang D, Liberek K, Yamamoto T, Georgopoulos C (1988) Initiation of lambda DNA replication reconstituted with purified lambda and *Escherichia coli* replication proteins. Biochim Biophys Acta 951:344–350

Zylicz M, King FW, Wawrzynow A (2001) Hsp70 interactions with the p53 tumour suppressor protein. EMBO J 20:4634–4638

Zylicz M, LeBowitz JH, McMacken R, Georgopoulos C (1983) The dnaK protein of *Escherichia coli* possesses an ATPase and autophosphorylating activity and is essential in an *in vitro* DNA replication system. Proc Natl Acad Sci USA 80:6431–6435

Zylicz M, Wawrzynow A, Marszalek J, Liberek K, Banecki B, Konieczny I, Blaszczak P, Jakóbkiewicz J, Gonciarz-Swiatek M, Duchniewicz M, Puzewicz J, Krzewska J (1999) Role of chaperones in replication of bacteriophage λ DNA. In: Bukau B (ed) Molecular chaperones and folding catalysts. Regulation, cellular function and mechanism. Harwood Academic Publishers, Amsterdam, pp 295–311

Rev Physiol Biochem Pharmacol (2005) 153:47–99
DOI 10.1007/s10254-004-0037-1

C. Campo · A. Mason · D. Maouyo · O. Olsen · D. Yoo · P. A. Welling

Molecular mechanisms of membrane polarity in renal epithelial cells

Published online: 17 November 2004

Abstract Exciting discoveries in the last decade have cast light onto the fundamental mechanisms that underlie polarized trafficking in epithelial cells. It is now clear that epithelial cell membrane asymmetry is achieved by a combination of intracellular sorting operations, vectorial delivery mechanisms and plasmalemma-specific fusion and retention processes. Several well-defined signals that specify polarized segregation, sorting, or retention processes have, now, been described in a number of proteins. The intracellular machineries that decode and act on these signals are beginning to be described. In addition, the nature of the molecules that associate with intracellular trafficking vesicles to coordinate polarized delivery, tethering, docking, and fusion are also becoming understood. Combined with direct visualization of polarized sorting processes with new technologies in live-cell fluorescent microscopy, new and surprising insights into these once-elusive trafficking processes are emerging. Here we provide a review of these recent advances within an historically relevant context.

Abbreviations *AEE:* Apical Early Endosomes · *AP:* Adaptins or clathrin-adaptor complexes · *ARF:* ADP-ribosylation factor, Ras-like small G-proteins involved in vesicular trafficking · *BFA:* Brefeldin A, a fungal metabolite that inhibits ARF-dependent attachment of coat proteins · *BEE:* Basolateral Early Endosomes · *CASK:* Calcium/calmodulin-dependent protein kinase, MAGUK protein, also known as Lin-2 · *CRE:* Common Recycling Endosome · *DRM:* Detergent Resistant Membrane, a biochemical hallmark of a raft · *ECM:* Extracellular Matrix · *EEA1:* Early Endosome Antigen 1, a Rab effector and a marker for the early endosome · *FYVE:* Zinc-binding domain named after the proteins that it is found in; targets proteins to membrane lipids via interaction with phosphatidylinositol-3-phosphate, "PI3P finger protein" · *GFP:* Green Fluorescent Protein · *GPI:* Glycosyl Phosphatidyl Inositol · *HA:* Influenza Hemagglutinin, a prototypical apical marker protein · *MDCK:* Madin Darby Kidney Cells, a polarized epithelial cell line · *Munc18–1:* Mammalian UNC18, C. ELEGANS, syntaxin-binding protein 1 · *LDLR:* Low-Density Lipoprotein Receptor, a prototypical basolateral marker ·

C. Campo · A. Mason · D. Maouyo · O. Olsen · D. Yoo · P. A. Welling (✉)
Department of Physiology, University of Maryland School of Medicine,
Baltimore, MD 21201, USA
e-mail: pwelling@umaryland.edu

LLC-PK1: A pig kidney polarized epithelial cell model · *L27:* A protein–protein interaction domain first identified in Lin-2 and Lin-7 · *Lin:* Genes that were first identified in genetic screens for abnormal cell *lin*eage in C. Elegans · *MAGUK:* Membrane Associated Guanlyate Kinases, a family of PDZ proteins · *MT:* Microtubule · *NSF:* *N*-ethyl maleimide-sensitive factor, a component of the fusion machinery · *pIgR:* Polymeric Immunoglobin Receptor For IgA, a prototypical protein for basolateral to apical transcytosis · *Par:* Partitioning Defective Proteins · *PDZ:* A protein–protein interaction domain found in scaffolding proteins · *PGTI:* Post-Golgi transport intermediates · *Rab:* A family of RAS-related small GTP-binding proteins that play important roles in the regulation of exocytotic and endocytotic pathway · *Ral:* A family of RAS-related small GTP-binding proteins that have been implicated in vesicle trafficking, cell morphology, and signalling · *TfR:* Transferrin receptor, a prototypical marker for basolateral membrane and CRE recycling · *SAC:* Subapical Compartment, also known as ARE · *Sec 6/8:* Components of the exocyst, but the term is often used to refer to the entire mammalian exocyst complex · *SNAP:* Soluble NSF Attachment Factor, component of fusion machinery · *SNARE:* Soluble N-Ethyl Maleimide-Sensitive Attachment Factor Receptors, components of fusion machinery · *t-SNARE:* SNARES on target membrane · *v-SNARE:* SNARES on vesicle membrane, also known as VAMP/synaptobrevin · *TGN:* Trans-Golgi Network · *VAMP:* Vesicle Associated Membrane Protein, also known as v-SNARE · *VSV G:* Vesicular Stomatitis Virus coat protein, used as a prototypical basolateral marker

Introduction

The asymmetric distribution of membrane proteins on distinct plasma membrane domains is a fundamental property of epithelial cells and a central underpinning of vectorial transport processes, signal transduction mechanisms in development, cell–matrix interactions and barrier functions. Certainly, the homeostatic transport of solutes, electrolytes, and water in the kidney are made possible by the polarized expression of specific transport proteins and regulatory molecules on two separate plasma membrane domains. Consider, for example, the reabsorption of salt in the distal nephron, an important process for the maintenance of blood pressure. Because sodium channels and Na/K ATPase are specifically expressed on opposite plasma membrane domains (apical membrane, facing the tubule lumen, and the basolateral membrane, facing the interstitium, respectively), the sodium transport activities of these molecules are effectively linked in series, allowing the directional movement of sodium from tubule lumen to interstitium. Epithelial polarization processes have been the target of extensive study over the last quarter century and the subject of several excellent reviews (Caplan 1997; Matter and Mellman 1994; Mostov et al. 2000; Yeaman et al. 1999). Exciting new discoveries in the field have begun to illuminate the molecular mechanisms that underpin polarized targeting and expression of membrane proteins in epithelial cells.

According to the present understanding, polarized protein expression is determined by a combination of intracellular sorting processes and plasmalemma domain-specific retention operations. The present model holds that directed sorting, trafficking, and retention are dictated by signals embedded within the structures of polarized proteins. Intracellular sorting signals are read, interpreted, and acted on by intracellular trafficking machinery to segregate and package these membrane proteins into specialized apical and basolateral membrane transport vesicles. Intracellular vesicles, containing segregated cargo, then have the capacity to be directly shipped, docked, and fused with the appropriate domain, guaranteeing that cargo arrives at the correct plasmalemma. Once delivered to a particular plasma domain, many of these proteins can then be effectively anchored at these polarized locales through retention signals, directing interactions with the cytoskeleton or other membrane associated proteins. Fundamental mechanistic insights into polarization processes have begun to emerge with the identification of trafficking and retention signals, and the intracellular machinery that acts on them. Here we provide a review of the field, highlighting recent advances within an historically relevant context.

Sorting pathways

An appreciation of the fundamental mechanisms that underlie polarized sorting processes in epithelial cells can be gained from the vantage of the intracellular targeting pathways that are followed by membrane proteins en route to their polarized plasmalemma locales. In fact, elucidating the sorting pathway pursued by a particular protein in a particular epithelial cell type provides valuable, if not essential, clues about the basic nature of the sorting compartments and the prevailing or dominant sorting mechanism. In principle, three different routing pathways may be pursued (Evans 1980).

First, and conceptually the simplest, newly synthesized proteins can follow a *direct* route to their final, apical, or basolateral membrane destination (Fig. 1). In these cases, membrane proteins are sorted soon after synthesis and prior to plasmalemma delivery and are therefore said to follow the biosynthetic pathway. Analogous to protein sorting within the classic secretory or exocytic pathway (Palade 1975) of nonpolarized cells, the trans-Golgi network (TGN, see Abbreviations) has been considered to be the principal sorting station of the biosynthetic pathway in epithelial cells. Recent studies suggest the involvement of endosomal sorting compartments (see the section entitled "Biosynthetic pathway" below). While polarized protein sorting in the biosynthetic pathway is initiated and driven by the segregation of apical and basolateral membrane-destined cargo within these intracellular compartments, it is also dependent on other mechanisms, insuring that the intracellular transport carriers are directly shipped and docked with the appropriate polarized membrane domain. In these cases, plasma-membrane domain-specific retention mechanisms are not necessarily required but they can increase the fidelity of the polarization process and control cell surface density.

Second, newly synthesized membrane proteins may be randomly sent to both plasma membrane domains and then either selectively retained, degraded, or relocated to their polarized destinations by postendocytic sorting processes and transcytosis. In this trafficking process, called the *random* pathway, polarized sorting must be achieved at the plasmalemma itself by selective retention or within the postendocytic pathway by segregation processes or by a combination of both.

Third, in some systems, polarized expression of membrane proteins may arise by a circuitous routing process, often termed *indirect* trafficking. In this case, newly synthesized proteins are initially targeted to one plasma membrane domain and then either selectively retained or selectively retrieved and reshipped to the opposite membrane domain. Many apical membrane proteins in hepatocytes follow this route (Bartles et al. 1987). Recent studies indicate that certain apical proteins that are dependent on lipid rafts may use this

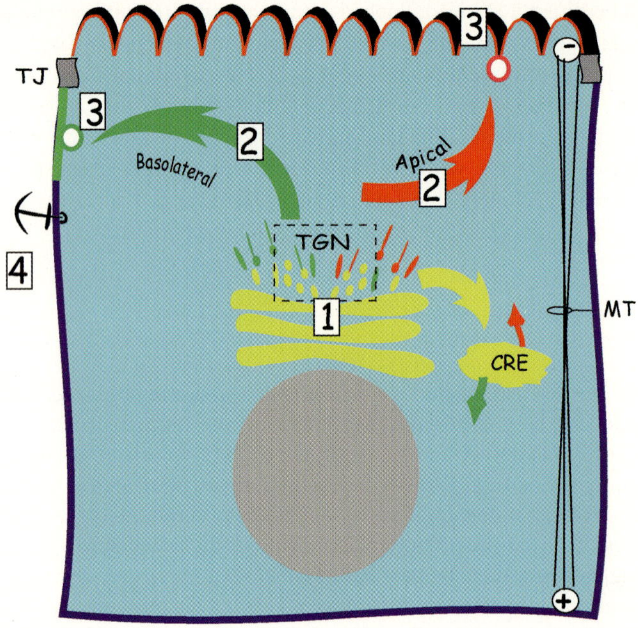

Fig. 1 Polarized expression of membrane proteins depends on hierarchical trafficking operations. Proteins traveling in the biosynthetic pathway are sequentially processed in as many as four different trafficking steps. (*1*) In this routing pathway, the sorting process is ultimately driven by the segregation of apical and basolateral membrane-destined proteins. This occurs soon after synthesis within the Trans Golgi network (*TGN*) and/or an endosomal sorting compartment (the common recycling endosome, *CRE*). (*2*) Following the biosynthetic sorting operation, apical and basolateral membrane-destined transport vesicles are directly shipped to the appropriate membrane domain, involving the cytoskeleton and the exocyst. (*3*) Specific membrane fusion machines guarantee that vesicles fuse with the appropriate target membrane, involving SNAREs and the exocyst. (*4*) Plasma-membrane domain-specific retention machinery, depicted as an anchor, can increase the fidelity of the polarization process by anchoring appropriately sorted proteins on the correct membrane domain. *TJ*, tight junction; *MT*, microtubules with plus ends (+) facing toward the apical membrane and minus ends (–) facing toward the basolateral membrane

pathway in renal epithelial cells, traveling first to the basolateral membrane and then being delivered to the apical surface by transcytosis (Polishchuk et al. 2004).

Polarized membrane protein expression in renal epithelia can be achieved by a combination of direct, random, and indirect routing mechanisms, depending on the membrane protein and type of epithelial cell. Consequently, segregation of polarized membrane proteins within biosynthetic and postendocytic sorting compartments as well as retention processes at specific polarized plasmalemma domains can contribute to membrane protein polarization in kidney cells.

Biosynthetic pathway

Proof of the well-traveled biosynthetic sorting pathway grew out of the development of a polarized renal epithelial cell model, Madin Darby Kidney Cells(MDCK; Leighton et al. 1969), and the seminal discovery that certain viruses emerge from infected MDCK cells in a polarized manner (Rodriguez-Boulan and Sabatini 1978). Certainly, the virus-infected MDCK model provided an early and extremely powerful tool to elucidate mechanisms of asymmetric sorting processes in epithelial cells. In this system, particular enveloped viruses, like the vesicular stomatitis virus (VSV), bud exclusively from the basolateral membrane while others, like the influenza virus, develop only from the apical membrane. The asymmetric budding response is a consequence of polarized expression of viral coat proteins. Steady-state localization of the VSV G coat protein occurs exclusively on the basolateral membrane. The Influenza hemagglutinin coat protein (HA), on the other hand, only accumulates on the apical membrane (Rodriguez-Boulan and Pendergast 1980). With the development of innovative plasmalemma domain-selective detection techniques, it was found that newly synthesized HA (Misek et al. 1984; Rodriguez-Boulan et al. 1984) and VSV G (Rindler et al. 1984) are rapidly and directly transported to their target membranes without any transitory stops on opposite plasma membrane domains. The result indicated that viral coat proteins must be sorted within an intracellular compartment and vectorially delivered to their target membrane, the *sine qua non* for a biosynthetic sorting process. Many later studies, employing similar pulse-chase and domain-selective detection techniques, such as surface biotinylation (Gottardi et al. 1995 for review), confirmed that endogenous membrane protein trafficking in MDCK cells (for example, Caplan et al. 1986) can also follow a direct sorting route to either the apical or basolateral membrane.

The specific location of a major biosynthetic sorting compartment was illuminated using biochemical and immuno-electron micrographic techniques to simultaneously track intracellular transit of apical and basolateral viral membrane proteins in MCDK cells doubly infected with pairs of viruses of opposite budding polarity. HA and VSV G proteins, en route to apical and basolateral membranes, were found to intermix at all stages of intracellular traffic from the endoplasmic reticulum through the Golgi cisternae. Thus, critical segregation steps must take place somewhere between the end of Golgi and the plasma membrane. This compartment has long been presumed to be the Trans-Golgi Network (TGN; Rindler et al. 1984). Support for segregation within the TGN came from observations that membrane protein traffic can be specifically blocked within the trans-Golgi by low-temperature (19–20°C) (Griffiths et al. 1985; Saraste and Kuismanen 1984). Upon release from block, viral membrane proteins proceed rapidly and directly to their appropriate plasma membrane without obligate stops in endosomes or other intracellular compartments (Griffiths et al. 1985; Pfeiffer et al. 1985; Matlin and Simons 1983). Distinct TGN vesicles (Bennett et al. 1988), separately containing apical and basolateral membrane cargo, can be resolved biochemically by immuno-isolation techniques after low-temperature block (Wandinger-Ness et al. 1990), providing strong evidence for a sorting event in the TGN. These classic studies have recently been corroborated by time-lapse fluorescence imaging studies of live cells transfected or infected with basolateral and apical membrane marker proteins. These studies have revealed that apical and basolateral membrane cargo progressively can segregate within domains of the Golgi and TGN, exclude resident proteins, and then exit in separate tubulovesicular carriers in direct route to the plasmalemma (Keller et al. 2001; Kreitzer et al. 2003).

Although there is a large body of data indicating that polarized trafficking from the TGN to the plasma membrane domains often proceeds in a direct fashion, it should be pointed out that sequential intracellular routing steps, involving intermediate endosomal compartments, have recently been identified. For instance, using an assay designed to measure the meeting of newly synthesized membrane proteins with endosomal compartments loaded with horseradish peroxidase, Orzech et al. (Orzech et al. 2000) found that the biosynthetic road traveled by polymeric immunoglobulin receptors can involve an endosomal compartment, most likely the common recycling endosome (CRE). Because apical and basolateral proteins intermix in this compartment (see the section entitled "Postendocytic pathway" below), the observation implies that the common endosome might also serve as a polarized sorting station for some proteins in the biosynthetic pathway. In fact, new work on the clathrin-adaptor complexes raises the possibility that biosynthetic sorting within the CRE, rather than the TGN, may be much more widespread than once believed (see section entitled "Adaptins"). At present it is unknown if trafficking from the TGN to the CRE occurs by default or is signal mediated. Likewise, the functional significance of multiple or parallel routing steps, involving the TGN and/or the CRE, in the biosynthetic pathways remains an open question. Orzech suggested that it may permit flexible regulation of sorting processes, possibly allowing cells to adjust targeting in response to changing physiological and developmental requirements (Orzech et al. 2000).

Whether sorting occurs in the TGN or CRE, the crucial, if not driving, step involves segregation of cargo and the assembly of trafficking vesicles by coat proteins and adaptor molecules or complexes. Selective attachment of specific coat factors to cargo is believed to promote segregation and concentration of select cargo while specifically marking vesicle carriers for either basolateral or apical membrane traffic. Selective portioning in specialized lipid domains may provide another mechanism (see section entitled "Rafts"). Structural evidence for distinct coat-based sorting domains in the TGN is provided by high-voltage electron microscopy studies combined with computer axial tomography. These studies reveal that the TGN consists of multiple tubules, containing clathrin or novel "lace-like" coats (Ladinsky et al. 1994, 1999, 2002). Moreover, Brefeldin A (BFA), a fungal metabolite that inhibits ADP-ribosylating factor (ARF)-dependent attachment of coat proteins, can cause missorting of proteins traveling in the biosynthetic pathway.

Postendocytic pathway

Precise sorting in the postendocytic pathway not only provides the chief polarizing mechanism for proteins that are processed in the indirect pathway by transcytosis (see Tuma and Hubbard 2003; Rojas and Apodaca 2002 for recent review of transcytosis), it is also required for the general maintenance of epithelial cell polarity. Certainly, because nearly 50% of a typical polarized plasma membrane is endocytosed per hour (Mellman 1996; Von Bonsdorff et al. 1985), specific sorting mechanisms must be in place to insure that endocytosed proteins are recycled back to the appropriate plasma membrane. Viewed another way, constant internalization and sorting in the postendocytic pathway may offer a continuous surveillance or proofreading mechanism, faithfully reshuffling proteins that happen to be randomly sorted or missorted in the biosynthetic pathway. Considering that sorting in the biosynthetic pathway only occurs once in the lifetime of a protein, continuous postendocytic sorting processes are quantitatively more significant.

Distinct intracellular sorting compartments have been identified in model epithelia, such as MDCK cells and CaCO-2 intestinal cells (Hughson and Hopkins 1990; Knight et al. 1995). As illuminated by studying intracellular trafficking itineraries of endocytotic recycling (e.g., transferrin and its receptor, TfR) and transcytotic marker proteins (e.g., IgA and its receptor, pIgR), the collective body of recent data is consistent with at least three, more likely four, distinct populations of endosomes (see Fig. 2). While generally reminiscent of the early and recycling endosomes that have been well described in nonpolarized

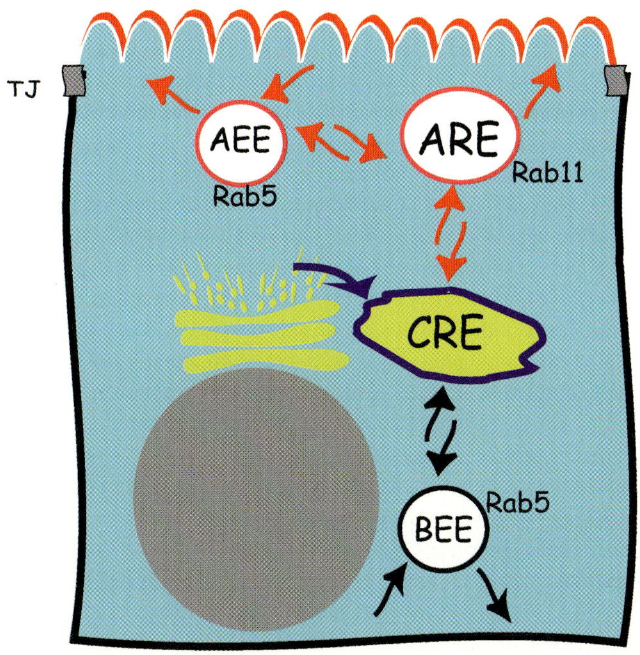

Fig. 2 Postendocytic sorting compartments in epithelial cells. Proteins internalized by clathrin-dependent pathways first accumulate in distinct populations of early endosomes found near the basolateral membrane (*BEE*, basolateral early endosome) or the apical membrane (*AEE*, apical early endosome). Early endosomes are can be identified by the presence of Rab 5 and/or EEA1. The common recycling endosome (*CRE*) receives endocytosed proteins from the apical membrane and basolateral membrane via the AEE (arrows not shown), the BEE, and possibly the ARE. The CRE also receives newly synthesized proteins from the TGN. Here, apical and basolateral membrane proteins are segregated and recycled back to the correct plasma membrane domain. The CRE is usually identified by its perinuclear location and the presence of Rab 11 or Rab 17, and an acidic pH. The ARE (*a*pical *re*cycling endosome, also known as SAC) is suspected to be an endocytic hub where apical membrane traffic is regulated. It is often identified by its apical location, the presence of Rab 11 and a neutral pH

cells (Mellman 1996), additional levels of complexity presumably reflect the added requirements for polarized sorting operations.

Material internalized by clathrin-dependent pathways from opposite poles of epithelial cells first accumulates with fluid-phase markers in spatially distinct populations of early endosomes found near the basolateral (BEE, Basolateral Early Endosomes) and apical membranes (AEE, Apical Early Endosomes) (Parton et al. 1989; Bomsel et al. 1989). While both of these peripheral compartments contain Rab 4 (Sheff et al. 1999) or Rab 5 (Bucci et al. 1994) and EEAI, a Rab effector (Wilson et al. 2000; Simonsen et al. 1998), the AEE and BEE appear to be biochemically and physiologically distinct from each other. AEE and BEE exhibit different overall protein compositions (Fialka et al. 1999), fusigenic properties (Fialka et al. 1999), recycling rates (Sheff et al. 1999), and dependence on filamentous actin (Sheff et al. 2002). Such observations presumably indicate that the AEE and BEE are separately designed for disparate sorting demands. Certainly, each population of early endosome supports rapid and efficient membrane protein recycling to its cognate plasma membrane, preventing mixing of apical and basolateral membrane proteins. Material endocytosed into AEE or BEE can also be separated from fluid-phase markers (Dunn et al. 1989) and be targeted into the lysosomal pathway or directed to a common recycling endosome (CRE) (reviewed by Mukherjee et al. 1997). Apical proteins endocytosed into the AEE may also travel through a specialized compartment, called the apical recycling endosome (ARE) (Leung et al. 2000).

The CRE, often identified by its perinuclear localization, the presence of Rab 11(Sheff et al. 1999) and/or Rab 17 (Hunziker and Peters 1998), absence of fluid phase markers, and a tubular morphology, is clearly distinct from early endosomes. Indeed, the CRE can be resolved from early endosomes physically, biochemically, and pharmacologically in MDCK cells (Sheff et al. 1999). Because the CRE receives and appropriately recycles internalized material from both the basolateral and the apical compartment (Bomsel et al. 1990; Parton et al. 1989), it is generally considered to be a major polarized sorting station. Recent work is beginning to provide a glimpse into a mechanism for some basolateral membrane proteins. Ultrastructural studies indicate that basolateral sorting of the transferrin receptor occurs in tubular extensions of the CRE (Odorizzi et al. 1996) on clathrin-studded vesicles (Futter et al. 1998). Reminiscent of the biosynthetic pathway, the sorting process in the CRE is sensitive to BFA (Matter et al. 1993; Wan et al. 1992; Wang et al. 2002; Futter et al. 1998). Because BFA does not induce gross disruption or fusion of the endocytic compartments in MDCK cells (Wang et al. 2002), it has been believed to alter endosomal sorting processes directly. Presumably by exerting its well-known inhibitory effect on the association of coat proteins (Kreis 1992), BFA prevents critical segregation steps in the CRE. Futter first presented evidence suggesting that γ-adaptin, a subunit of AP-1A and AP-1B clathrin adaptor complexes, marks clathrin-coated vesicles emanating from the tubular extensions of the CRE as basolateral membrane-destined carriers (Futter et al. 1998). More recent work indicates that direct interaction of cargo in the CRE with the AP-1B complex is sufficient to target proteins to the basolateral membrane (see section entitled “Adaptin”).

A three compartment model, comprised of the BEE, AEE, and CRE may be sufficient to describe the kinetic behaviors of the pIgR bound with IgA in the transcytotic pathway and the TfR in the recycling pathway of MDCK cells (Sheff et al. 1999, 2002). Nevertheless, a fourth compartment is likely to serve as a depot for membrane proteins traveling between the apical membrane and CRE (reviewed in Van Ijzendoorn et al. 2000; see Fig. 2), perhaps via the AEE (Leung et al. 2000). This compartment, called the apical recycling endosome (ARE) or the subapical compartment (SAC), has been the subject of great interest because it is suspected to be a postendocytic hub where apical membrane traffic is regulated to control apical membrane protein surface density (Van Ijzendoorn et al. 2000). Nevertheless, efforts to differentiate the ARE from apical-membrane transport intermediates or, more importantly, the CRE, have been hampered because of their overlapping or common properties. Indeed, both ARE and CRE compartments are located in a subapical, perinuclear, or pericentriolar region of the cell, are devoid of fluid-phase markers, are dependent on microtubules, and many contain a similar complement of Rab proteins, including Rab 11, Rab 17, and Rab 25 (Casanova et al. 1999; Sheff et al. 1999; Goldenring et al. 1996; Hunziker and Peters 1998). More recent studies (Brown et al. 2000; Wang et al. 2000; Leung et al. 2000) provide several criteria for distinguishing ARE from the CRE in MDCK cells. Brown defined the ARE as an endocytic compartment at the extreme apical pole that is enriched in apical proteins. The CRE, in contrast, distributes throughout the apical and lateral cytoplasm and contains apical and basolateral proteins in equal proportion. Furthermore, as resolved by confocal microscopy, Rab 11 predominately associates with the apical-membrane-protein-enriched vesicles at the extreme apical pole rather than with the CRE. A sharp demarcation in the intravesicular pH of the CRE (acidic) and ARE (neutral) provides a functional dissimilarity, which can be used to define the two compartments (Wang et al. 2000).

How polarized sorting is achieved in the endocytotic pathway remains a fundamental question. The identification of various endocytic compartments in epithelial cells provides a road map to find key sorting locales. The notion that postendocytic sorting may share fundamental mechanisms with the biosynthetic pathway provides an obvious, if not oversimplified, hypothesis to test.

Sorting signals

According to our present understanding, polarized sorting processes are signal dependent. That is, signals embedded within the structure of polarized membrane proteins contain cell localization information. These signals operate as the molecular equivalents of zip codes. They are read, interpreted, and acted on by intracellular sorting machineries, which in turn shuffle, retain, or retrieve molecules to appropriate membrane domains. A great deal of research activity has focused on identifying these signals to help illuminate the mechanisms of polarized sorting operations.

Generally, polarized sorting determinants are identified in mutagenesis studies as structural elements necessary for polarized membrane expression. By definition, removal of such a structure will alter polarized expression. The type of missorting phenotype that is produced, however, depends on the presence or absence of other sorting signals contained within the polarized membrane protein. When the protein is devoid of any other sorting information, ablation of the sorting signal will result in a nonpolarized expression pattern. In contrast, removal of a dominant polarized sorting signal can cause preferential localization to the opposite membrane domain when recessive sorting signals are present.

Once a particular structure is recognized as a necessary polarized sorting determinant, it is instructive to determine whether it is sufficient to act in an autonomous fashion and direct a polarized sorting operation in isolation. Such a test usually involves transplanting the structure of interest onto a reporter protein that is otherwise expressed in a nonpolarized fashion and examining whether it confers a polarized sorting phenotype. Combined with further mutagenesis, such studies may reveal the minimum and precise sequence requirements for a polarized sorting operation, and thereby define the sorting signal signature.

Basolateral sorting signals

Two different classes of autonomous basolateral sorting signals have been defined (Matter and Mellman 1994) from work on unrelated proteins. One class is similar to clathrin-dependent endocytic or lysosomal sorting signals (see Bonifacino and Traub 2003 for review). These include the basolateral sorting signals, typified by short tyrosine-containing motifs (NPXY or YXXØ, where Φ is a hydrophobic residue; Bonifacino and Dell'Angelica 1999). These are seen, for example, in the LDL receptor proximal signal (Matter et al.

Table 1 Three binding classes of PDZ domains. Ligands of different classes of PDZ domains are largely distinguished by residues at the O and –2 positions, where O denotes the extreme COOH terminal residue (*). Subcellular localization is not generalized; it is listed for the specific PDZ protein/ligand example in the corresponding row

PDZ binding class	Canonical ligand	Example (ligand/PDZ protein)	Localization
Class I	X-(S/T)-X-ϕ-*	DSAI* (Kir 2.3)/Lin-7 (Olsen et al. 2001)	Basolateral
		DTRL* (CFTR)/NHERF (Moyer et al. 2000)	Apical
Class II	X-(ϕ/Ψ)-X-ϕ-*	EFYA* (Syndecan-2)CASK (Daniels et al. 1998)	Basolateral
Class III	X-(D/E)-X-ϕ-* or (E/D)-X-W-C/S	SEPL*KIF17/Mint1(lin-2) (Setou et al. 2000) (Maximov et al. 1999)	Trafficking vesicles

ϕ, Hydrophobic; Ψ, aromatic; *, COOH terminus.

1992), the VSV-G protein (Thomas and Roth 1994; Thomas et al. 1993), and the asialoglycoprotein receptor (Geffen et al. 1993). Alternatively, dileucine/dihydrophobic signals have been identified in the IgFc receptor (Hunziker and Fumey 1994; Matter et al. 1994), the MHC, class II protein (Simonsen et al. 1999b), the CD44 adhesion factor (Sheikh and Isacke 1996), E-cadherin (Miranda et al. 2004), and the Norepinephrine transporter (Gu et al. 2001). Present evidence suggests that these signals may direct interaction with clathrin-adaptor molecules (see section entitled “Adaptin”).

The other class of basolateral sorting signals, distinguished only by being unrelated to endocytic signaling motifs, is likely to represent several distinct kinds of sorting signals (Matter and Mellman 1994; Le Gall et al. 1997; Reich et al. 1996; Simmen et al. 1999). These include acidic cluster motifs that are either juxtaposed to critical di-hydrophobic residues [e.g., in furin (Simmen et al. 1999) and the stem cell factor (Wehrle-Haller and Imhof 2001)] or penultimate tyrosine residues (e.g., the LDL receptor distal determinant; Matter et al. 1992). The H/RXXV motif in the polymeric immunoglobulin receptor, pIgR, (Casanova et al. 1991; Aroeti et al. 1993) has been proposed to form a beta-turn structure that interacts with BFA-sensitive adaptor complexes (Reich et al. 1996) to direct basolateral sorting. Other basolateral sorting signals, like those in the Kir 2.3 channel (Le Maout et al. 2001), the GAT-2 GABA transporter (Brown et al. 2004), and betaine transporters (Perego et al. 1997), are found at the extreme carboxyl terminus and juxtapose PDZ binding motifs (see section entitled “PDZ proteins” and Table 1). These signals appear to operate independently of the PDZ ligand. Interestingly, part of the basolateral sorting determinant in Kir 2.3 shares some resemblance to the more precisely defined basolateral sorting signal in the GAT-2 GABA transporter (defined by the sequence, LRLTELE), raising the possibility of a distinct class of conserved basolateral sorting signals. Still other unrelated and unique autonomous basolateral sorting signals have been described in transferrin (Odorizzi and Trowbridge 1997) and the neural cell adhesion molecules (N-CAM) (Le Gall et al. 1997).

Apical sorting signals

Apical sorting signals are less well defined and, until recently, were believed to reside almost exclusively within membrane domains or in extracellular regions. For instance, GPI-linkage and certain transmembrane domain structures, which allow selective partitioning into lipid rafts, have been implicated as apical sorting determinants in some apically expressed proteins (see section entitled "Rafts"). Other transmembrane domain structures drive apical membrane targeting in a manner that is not dependent on raft compartmentalization, such as in the case of the gastric H/K ATPase alpha subunit (Dunbar et al. 2000). In other cases, N- and O-linked glycosylation appears to be necessary for correct apical delivery (Scheiffele et al. 1995; Wagner et al. 1995; Yeaman et al. 1997; Jacob et al. 2000; Rodriguez-Boulan and Gonzalez 1999; Potter et al. 2004). This may reflect the requirement for proper cargo folding (Rodriguez-Boulan and Gonzalez 1999) and/or the involvement of apical cargo receptor lectins, such as VIP36 (Hara-Kuge et al. 2002).

More recently, apical sorting determinants have been identified in cytoplasmic domains. For instance, PDZ domain binding motifs in GAT-3, a GABA transporter (Muth et al. 1998), the sodium-phosphate transporter, NaPi II (Hernando et al. 2002), and the Cystic Fibrosis Transmembrane Regulator, CFTR (Moyer et al. 2000), appear to be required for efficient apical localization of these molecules. The cytoplasmic tail of rhodopsin contains an unrelated apical sorting determinant, which directly interacts with microtubule motor proteins to direct apical targeting (Tai et al. 2001; Chuang and Sung 1998). Interestingly, the cytoplasmic tail of the main endocytic receptor of the proximal tubule, megalin, has recently been shown to possess apical sorting information that contains an NPXY-type motif (Takeda et al. 2003), raising the surprising and fascinating possibility that some apical sorting operations may also involve interactions with clathrin-adaptor molecules (Nagai et al. 2003)

Sorting signal recognition

A great deal of research activity has focused on identifying and characterizing the intracellular machinery that directly interacts with and decodes intracellular sorting and plasma membrane retention signals. Work in this area has largely proceeded along two experimental fronts. Candidates for sorting machinery have been discovered using biochemical and genetic methods to capture proteins that directly interact with sorting signals. Others have been identified using in silico analysis to mine the genome for epithelial-specific paralogs of known trafficking proteins. Once identified, candidate proteins are tested for the hallmarks of signal-dependent trafficking operations. Ideally, these include: (1) a similar sequence requirement for cargo-interaction and sorting functions, (2) localization to the appropriate sorting compartment, (3) a similar missorting phenotype should be observed whether the target sorting signal is mutated or the interaction with cargo is disrupted in vivo. While few of the presently identified sorting machinery candidates meet all of these rigorous standards, their discovery has greatly advanced our mechanistic understanding of polarization processes. Below we discuss our present understanding of these sorting machinery candidates.

Clathrin adaptors

The resemblance of tyrosine and dileucine-based basolateral sorting signals to well-established, clathrin-dependent endosomal and lysosomal targeting signals (see Bonifacino and Traub 2003 for review) has long suggested the possibility that basolateral membrane sorting processes might employ a related mechanism (Matter and Mellman 1994). The breakthrough discovery and characterization of a novel, epithelial-specific "medium" subunit of the clathrin adaptor AP-1 complex, called μ1B (Ohno et al. 1999), coupled with recent studies on the AP-4 clathrin adaptor complex (Simmen et al. 2002) strongly suggest that this is, in fact, the case.

Tyrosine- and dileucine-based endosomal and lysosomal sorting signals have been well known to act as target-recognition sites for specific subunits of heterotetrameric clathrin adaptor complexes (AP) called adaptins (reviewed in Bonifacino and Dell'Angelica 1999). Four adaptor complexes, consisting of two large subunits (γ, α, δ, ϵ, and β1–4), one medium subunit (μ11–4), and one small subunit (σ 1–4), have been described [AP-1 (β1, γ, μ1A or μ1B, σ 1); AP-2 (β2, α, μ2, σ2); AP-3 (β3, δ, μ3, σ3), or AP-4 (β4, ϵ, μ4, σ4) (Boehm and Bonifacino 2001)]. Medium subunits (μ1-μ4) of the clathrin adaptor complex specifically bind to the common endocytotic "YXXØ" sorting motif (see section entitled "Sorting signals") (Ohno et al. 1995, 1996) through a structurally-defined interaction often described as "a two-pinned plug fitting into a socket" (Owen and Evans 1998). The larger β subunits, on the other hand, have been shown to recognize dileucine motifs (Rapoport et al. 1998). Since clathrin adaptor complexes also participate in the formation of coated vesicles, their interaction with membrane proteins carrying these signals effectively marks them as cargo for inclusion into vesicles. Depending on the type of AP complex interaction, these vesicles are targeted to different destinations. For instance, the AP-1 complex is responsible for the delivery of proteins from the trans-Golgi to the endosomal-lysosomal system, while AP-2 mediates rapid internalization from the plasma membrane, and AP-3 has been implicated in alternative pathways to endosomes or lysosomes (Robinson and Bonifacino 2001). The subsets of YXXØ signals recognized by μ1, μ2, μ3, and μ4 overlap to a significant extent. Nevertheless, each chain exhibits differing preferences for residues neighboring the critical tyrosine residue (Ohno et al. 1998), presumably providing a basis for sorting selectivity by the YXXØ signal.

AP-1B

The first experimental evidence that the μ1B adaptin is actually involved in a basolateral sorting event, came from a series of elegant studies (Folsch et al. 1999) utilizing a pig kidney epithelial cell line called LLC-PK1 (Hull et al. 1976). Curiously, these cells do not express detectable amounts of μ1B (Ohno et al. 1999). Moreover, Roush et al. had previously shown that several proteins, which depend on tyrosine-based sorting signals for basolateral membrane localization in MDCK cells, are processed to the apical membrane in LLC-PK1 cells (Roush et al. 1998). Such a result might be predicted if μ1B was a crucial component of a basolateral sorting machine. Following these observations, Folsch et al. developed a LLC-PK1 cell line that stably expressed μ1B and tested whether expression of this adaptin was sufficient to rescue the apparent basolateral-sorting defect. They found that μ1B assembled with other AP-1 components, forming a complex, AP-1B, that is biochemically and spatially distinct from AP-1A (Folsch et al. 2003). Remarkably, both the LDL and transferrin receptors, which are mislocalized to the apical membrane in wild-type LLC-PK1cells, are appropriately expressed on the basolateral membrane in LLC-PK1 cells transfected with μ1B. These observations provided convincing evidence for a basolateral sorting process that is absolutely dependent on the AP-1 complex containing μ1B.

The most likely explanation for the requirement of the epithelial-specific adaptin, μ1B, is that it directly interacts with tyrosine-based basolateral membrane sorting signals. In such a scheme, proteins carrying these signals would be selected for inclusion into AP1B-dependent vesicles that are destined for the basolateral membrane. Recent biochemical and immunocytochemical studies support this notion, at least for the LDL receptor (LDLR) (Folsch et al. 2001). The LDLR can be coimmunoprecipitated with the entire AP-1B complex, and as measured by yeast two-hybrid, can directly interact, albeit weakly, with μ1B (Folsch et al. 2001). Moreover, as expected for a bona fide signal recognition event in the biosynthetic pathway, early immunofluorescence and electron microscopic localization of epitope-tagged μ1B suggested that AP-1B interaction with LDLR occurs within the TGN on clathrin-coated buds and vesicles (Folsch et al. 2001). Later studies indicated that AP-1B complexes are confined to perinuclear regions close to the TGN and colocalize with endocytosed transferrin, typical of common recycling endosomes (CRE) (Ang et al. 2003; Folsch et al. 2003).

Surprising results of studies designed to determine the pathway where AP-1B operates turned the spotlight onto the CRE as a major, perhaps common, polarized sorting station. These recent observations indicate that newly synthesized AP-1B-dependent cargo, such as VSV-G, can colocalize with AP-1B in transferrin-positive compartments immediately upon exit from the Golgi (Folsch et al. 2003). This strongly suggests that AP-1B performs sorting functions for proteins traveling in the biosynthetic pathway at the CRE rather than at the TGN, as originally believed (see section entitled "Biosynthetic pathway"). By contrast, careful examination of LDLR trafficking in LLCPK1 cells revealed a different sorting process and uncovered a role for μ1B in the postendocytic pathway (Gan et al. 2002). Gan et al. discovered that LDLR are directly targeted to the basolateral membrane by a μ1B -independent process. Amazingly, however, steady state expression of LDLR on the basolateral membrane still relies on AP1-B. This occurs because AP-1B also apparently functions to appropriately recycle internalized receptors back to the basolateral membrane. Without it, endocytosed receptors are missorted to the apical membrane. As first suggested by Traub and Apodaca (2003), the seemingly disparate results of Folsch et al. and Gan et al. can be reconciled by a hybrid model in which AP-1B operates in both biosynthetic and postendocytic pathways, depending on the cargo involved.

AP-1B appears to be functionally coupled to other basolateral sorting machinery. For instance, Cdc42, a Rho GTPase previously implicated in basolateral transport (Kroschewski et al. 1999; Musch et al. 2001), as well as Rab 8 and the exocyst (see sections entitled "Rabs" and "Docking and fusion," below) have recently been shown to operate in the AP-1B sorting pathway (Ang et al. 2003; Folsch et al. 2003). This potentially links AP-1B-dependent cargo selection and vesicle formation to a specific basolateral membrane-domain docking and fusion process.

It is clear that μ1B is a component of a basolateral membrane sorting machine. Nevertheless, several questions remain unanswered, particularly regarding mechanisms of μ1B subunit interactions with basolateral membrane-destined cargo and how these interactions decode the sorting signals. This issue is particularly relevant for μ1B-dependent basolateral proteins, such as the LDLR, which contain tyrosine-based basolateral sorting signals that do not conform to the canonical "YXXØ " adaptin-binding motif (Matter et al. 1992). Other μ1B -dependent proteins, such as the TfR, rely on sorting signals that share absolutely no resemblance to known clathrin-adaptor binding motifs (see section entitled "Sorting signals"). A simple and straightforward answer to this puzzle is that some basolateral sorting determinants might interact directly with μ1B through a distinct binding site other than that used by YXXØ motifs. In support of this idea, mutant μ1B subunits, rendered defective for YXXØ binding, are able to support basolateral membrane expression of LDLR and TfR but are unable to support basolateral membrane targeting of those proteins that require a canonical tyrosine motif for basolateral sorting (Sugimoto et al. 2002). While highly suggestive, it should be pointed out that the mechanisms by which these noncanonical sorting signals bind to μ1B have not been characterized. Thus, it is impossible to judge if direct μ1B interaction is sufficient to decode these types of sorting signals.

In the absence of definitive interaction data, other mechanisms must also be considered. Certainly, given the broad range of proteins with different basolateral sorting signals that depend on μ1B for basolateral membrane expression, it is tempting to speculate that μ1B may also carry out indirect, but essential, basolateral sorting functions. Consistent with this idea, Folsch et al. found that μ1B expression also greatly improved the overall monolayer organization of LLC-PK1 cells and speculated that μ1B might induce the expression of one or more endogenous proteins, like integrins (see section entitled "Genesis of polarity," below) that play important roles in epithelial morphogenesis and the induction of epithelial polarity (Folsch et al. 1999).

It should be emphasized that AP-1B-dependent cargo-selection does not provide a universal mechanism for basolateral sorting in the biosynthetic pathway. Certainly, a number of proteins, such as the IgG Fc recptor FcR11-B2 (Roush et al. 1998) and the Kir 2.3 channel (Le Maout et al. 2001) are directly routed and stably expressed on the basolateral membrane of wild-type LLC-PK1 cells, implying an AP-1B-independent sorting process. The FcR11-B2 contains a dileucine-type basolateral targeting signal (Roush et al. 1998), which presumably has the capacity to interact with β-adaptin subunits (Rapoport et al. 1998). Thus, other adaptin complexes (such as AP-4, see below) or adaptin-like molecules (Zhu et al. 2001; Puertollano et al. 2001) may mediate basolateral targeting of molecules containing these types of sorting signals.

AP-4

Simmen et al. recently provided evidence for the involvement of the AP-4 adaptor complex in basolateral sorting (Simmen et al. 2002). They found that AP-4 localizes in the TGN and endosomes and showed that cytoplasmic MDCK cell fractions, enriched in the AP-4 complex, could interact with a diverse set of basolateral sorting signals in vitro, including those found in furin (Simmen et al. 1999), LDLR (Matter et al. 1993), and the TfR (Odorizzi and Trowbridge 1997) (see section entitled "Sorting signals"). RNA antisense-mediated knockdown of the medium AP-4 subunit, μ4, in MDCK cells produced an apical missorting phenotype of furin and LDLR, but not transferrin. The results suggest that AP-4 interaction with some target proteins may actually drive a basolateral sorting process. It will be important to determine how AP-4 interacts with these sorting signals and to ascertain whether these interactions are sufficient to decode basolateral sorting instructions.

Rabs

Members of the Rab family of small GTPases regulate the spatial and temporal organization of numerous membrane trafficking processes, including vesicle budding, targeting, and tethering as well as the docking and priming stages of vesicle fusion (reviewed in Schimmoller et al. 1998). Because they have highly specific localizations within different sorting stations, Rabs are considered excellent candidates for coordinators of specific trafficking events in epithelial cells (reviewed in Stow 1995). In fact, accumulating evidence strongly suggests that several Rabs are important components of polarized sorting machines. While Rabs have traditionally been considered to only interact indirectly with vesicular cargo by common vesicular affinities, recent studies reveal that cargo proteins can be direct effectors of Rabs. In these few cases, specific Rabs appear to directly interact with sorting signals.

Several Rabs have been implicated in apical membrane sorting processes. For instance, the epithelial-specific Rabs, Rab 17 (Lutcke et al. 1993; Zacchi et al. 1998) and Rab 25 (Goldenring et al. 1993), along with the more widely expressed Rab 11a, localize within discrete populations of apical endosomes (Casanova et al. 1999). Using an inducible system to express wild-type, dominant negative, or constitutively active Rab mutants, Wang et al. showed that Rab 11a and 25 are required for selective postendocytic vesicular transport steps that control apical rather than basolateral recycling processes in MDCK cells (Wang et al. 2000). Interestingly, Rab 11a redistributes to apical secretory cannaliculi during stimulation of acid secretion in gastric parietal cells (Calhoun et al. 1998). Conditional expression of a dominant negative mutant Rab 11a is capable of specifically inhibiting regulated translocation of the H/K-ATPase from the tubulovesicular network to the apical cannulicular membrane (Duman et al. 1999). Similar studies implicated Rab 17 in the regulation of traffic through the subapical population of endosomes. These endosomes are considered to be ARE in MDCK (Hunziker and Peters 1998) and Eph mammary epithelial cells (Zacchi et al. 1998). Thus, three Rab proteins, Rab 11a, Rab 17, and Rab 25, appear to regulate trafficking into or out of discrete endosomal systems, which, in turn, control apical targeting of transport vesicles.

Other Rabs, particularly Rab 8, have been implicated in basolateral membrane processing steps (Huber et al. 1993; Ang et al. 2003). In MDCK cells, Rab 8 localizes within or near the Golgi complex and within a population of basolateral membrane-destined vesicles that carry VSV-G as cargo. Huber et al. (1993) showed that a peptide derived from the hypervariable COOH-terminal region of Rab 8 could selectively inhibit biosynthetic membrane traffic of VSV-G from the TGN to the basolateral plasma membrane without affecting influenza HA traffic from the TGN to the apical surface. More recent studies indicate that expression of a constitutively activated GTP hydrolysis mutant of Rab 8 selectively inhibits basolateral transport of newly synthesized membrane proteins traveling in the μ1B adaptor-dependent pathway. Thus, Rab 8 participates in biosynthetic and postendocytic sorting at the CRE (Ang et al. 2003).

While many of the precise details of Rab function are still hazy, accumulating evidence has begun to provide a plausible generalized mechanistic model (Schimmoller et al. 1998). It is presently believed that Rab proteins orchestrate trafficking events by specifically associating with a precise membrane compartment and then recruiting a unique set of effector proteins to that compartment, depending on the particular Rab and its GTP/GDP-dependent conformational state. Perhaps the best understood Rab effectors are tethering molecules, like those in the exocyst complex (see section entitled "Exocyst"). Certainly, genetic and biochemical experiments in yeast revealed that the active, GTP-bound form of the prototypical Rab, Sec 4p, directly interacts with a component of the exocyst, Sec 15p, to initiate the assembly of the entire exocyst complex in yeast (Guo et al. 1999). Not surprisingly, tethering proteins also act as Rab effectors in epithelia. For instance, the most likely ortholog of Sec 4 is Rab 8, the Rab implicated in basolateral sorting (see above). Moreover Rab 5, a common component of the apical and basolateral early endosomes (Simonsen et al. 1998), interacts with the FYVE-finger protein, EEAI, (Simonsen et al. 1998; Wilson et al. 2000) to confer targeting specificity before SNARE-dependent early endosomal fusion events (Christoforidis et al. 1999; Simonsen et al. 1999a; McBride et al. 1999). More germane to polarized targeting, Rip 11, an epithelial-specific Rab 11 effector, appears to tether vesicles emanating from the apical recycling endosome en route to the apical membrane (Prekeris et al. 2000). The identification of other types of Rab effectors in nonpolarized cells, such as vesicle coat proteins or adaptors (Carroll et al. 2001) and specific microtubule- or actin-based motors (reviewed in Hammer and Wu 2002), provides intriguing candidates for vesicle formation and directed trafficking in polarized epithelial cells.

Each Rab protein can potentially act in multiple sorting roles. However, transport specificity is achieved because their performances are likely to be set on only one, or a limited number, of specific sorting stages. It has been generally assumed that Rabs are recruited to specific membrane domains as parts of trafficking complexes, connected only indirectly to vesicular cargo by common vesicular affinities or by adaptor molecules and coat proteins. A recent study from K. Mostov's group (Van Ijzendoorn et al. 2002), revealing that Rab 3b directly interacts with the polymeric immunoglobulin receptor, now suggests that certain Rabs directly bind to specific cargo proteins to affect their trafficking itinerary.

Van Ijzendoorn et al. discovered that Rab 3b, a member of a Rab subfamily well known to control regulated exocytic events, directly interacts with the polymeric immunoglobulin receptor, pIgR, within vesicular structures near the apical surface of MDCK cells. They showed that the active, GTP-bound form of Rab 3b binds to a specific 14 amino-acid site in the cytoplasmic domain of the receptor that was previously shown to be required for pIgR transcytosis, implicating the interaction in a transcytotic trafficking event. Using a combination of biochemical, physiological, and cell biological analyses, the group found that regulated dissociation of Rab 3b from pIgR stimulates basolateral to apical pIgR transcytosis, presumably by triggering translocation of receptor-containing-vesicles from the ARE to the apical membrane. Collectively, the study provides strong evidence for the notion that direct interaction of vesicle cargo with a Rab can determine its own intracellular traffic. It remains to be determined how widespread the phenomenon is. As pointed out by Van Ijzendoorn, the direct cargo interaction mechanism may have evolved for very abundant membrane proteins, like pIgR, which might require a dedicated and "private" trafficking mechanism to maintain a high fidelity of robust intracellular trafficking.

PDZ proteins

PDZ domains, named after the homologous group of proteins that they were originally identified [**P**SD 95, a postsynaptic density protein (Hunt et al. 1996), **D**lg (Drosophila **D**isc large tumor suppressor), and **Z**O-1 (zona occludens, the tight junction protein) (Jesaitis and Goodenough 1994)], are ~ 90 amino acid protein-interaction modules that bind short protein motifs (four amino-acid) generally (Songyang et al. 1997), but not always (Hillier et al. 1999; Harris et al. 2001), found at the extreme COOH-terminus of target proteins (Table 1). PDZ domains have now been identified in a variety of proteins. Most of these molecules contain multiple PDZ domains or other protein–protein interaction domains, allowing them to act as molecular scaffolds that facilitate multiprotein complex formation and organize expression of target proteins on particular membrane domains (Gomperts 1996). Obviously, such functions are well suited for polarized sorting and retention operations. An ever-growing body of work has strongly implicated PDZ proteins in targeting and clustering various receptors, channels, transporters, and signal transduction elements at specific plasma membrane domains in neurons (Sheng and Sala 2001; Gomperts 1996), muscle (Adams et al. 2001), and the *Drosophila* visual system (Xu et al. 1998). Concomitantly, a number of studies also indicate important roles for PDZ proteins in polarized epithelial sorting processes.

Significant, albeit correlative, clues about PDZ protein function in epithelia are provided by their curious localization. Certainly, with growing precedent in other systems, observations that various PDZ proteins are preferentially expressed at polarized membrane domains or within critical sorting compartments indicate a capacity to perform polarized retention or sorting operations. There are examples of PDZ proteins that predominately reside at the basolateral membrane of certain intestinal and renal epithelia. These include syntrophin (Kachinsky et al. 1999), Lin-7 (Straight et al. 2000), the ErbB interacting protein, ERBIN, (Borg et al. 2000), and certain members of the MAGUK (Membrane Associated Guanlyate Kinase) family of PDZ proteins (Anderson 1996), such as CASK (Cohen et al. 1998), PSD-93 (Tojo et al. 1999), and SAP97 [also known as Discs large homolog 1 (Wu et al. 1998)]. Other PDZ proteins, like the Sodium Hydrogen Exchange Regulator Factors [also known as NHERF 1 or EBP-50 and NHERF 2, E3-KARP (Shenolikar and Weinman 2001; Wade et al. 2001], PSD-95 (Tojo et al. 1999), NaPi-Cap1/2 [also known as PDKZ1, Cap 70 or CLAMP (Gisler et al. 2001)], and IKEPP [intestinal and kidney-enriched PDZ protein (Scott et al. 2002)] are chiefly expressed on or near the apical membrane. Still others, like CAL (CFTR-associated ligand), which is primarily located in the Golgi (Cheng et al. 2002) or syntenin, which is found in apical recycling endosomes (Fialka et al. 1999), reside in biosynthetic or endocytotic sorting compartments.

As would be predicted if PDZ proteins directly affect polarized sorting or retention of proteins that interact with them, a PDZ binding motif appears to be necessary for polarized expression of a variety of different membrane proteins in epithelial cells. Working to identify polarized sorting signals in the GABA transporters or GATs (Ahn et al. 1996), Muth et al. (Muth et al. 1998) provided one of the first examples. These investigators found that deletion of a COOH-terminal sequence resembling a PDZ binding motif from the apically expressed isoform, GAT-3, caused the transporter to randomly localize to both apical and basolateral membranes. Likewise, apical expression of the Cystic Fibrosis Transmembrane Regulator, CFTR, (Moyer et al. 2000; Benharouga et al. 2003) and the sodium-phosphate cotransporter (Karim-Jimenez et al. 2001; Hernando et al. 2001), which have the capacity to interact with the apical PDZ proteins NHERF and NaPi-Cap2 (Wang et al. 1998, 2001; Short et al. 1998; Karim-Jimenez et al. 2001), require an intact PDZ binding motif for apical membrane localization. In the case of CFTR, the PDZ binding motif appears to be read in the postendocytotic pathway to coordinate efficient expression on the apical membrane (Moyer et al. 2000). Basolateral membrane expression of several membrane proteins has also been found to require a PDZ binding motif. For instance,

ErbB receptors, which play crucial roles in morphogenesis and oncogenesis, interact with a basolateral PDZ protein, called ERBIN, and require a PDZ binding motif for basolateral membrane expression (Borg et al. 2000). Efficient basolateral membrane expression of the inwardly rectifying potassium channel, Kir 2.3, (Olsen et al. 2002) and γ-amino butyric acid transporter, BGT-1 (Perego et al. 1999), both of which interact with the basolateral PDZ protein Lin-7, also require an intact PDZ binding site.

In principle, PDZ proteins could coordinate polarized expression of target proteins by all imaginable mechanisms, involving any level of the sorting process. Certainly in nonepithelial cells, different PDZ proteins have been shown to mediate anterograde trafficking in the biosynthetic pathway (Fernandez-Larrea et al. 1999; Scott et al. 2001; Standley et al. 2000), control delivery (Setou et al. 2000) to and retention on the plasmalemma (Kim et al. 1995), and play roles in postendocytotic recycling (Cao et al. 1999). In polarized epithelial models, present evidence strongly supports the idea that PDZ–protein interactions orchestrate plasma membrane retention and postendocytic sorting operations. However, because of the large number of different PDZ proteins that can potentially interact with any one target, it has been a challenge to ascribe a particular PDZ protein to a specific polarized trafficking operation. With the advent of new gene silencing methods and PDZ protein knockout models, such as the one used to explore NHERF-1 function (Weinman et al. 2003; Shenolikar et al. 2002), the precise roles of PDZ proteins should begin to be illuminated. In the meantime, the rich history of the Lin-7 PDZ protein complex in *C. Elegans* along with its vertebrate orthologs provides an example of the mechanisms by which PDZ proteins coordinate polarized expression of target proteins in epithelial cells.

Lin-7 system

Perhaps the first definitive evidence that a specific PDZ protein can coordinate the polarized targeting of its interacting protein partner evolved from the identification of three different PDZ-protein encoding genes, *lin-7, lin-2, and lin-10* (Lin for abnormal cell *line*-age), required for vulva progenitor cell (VPC) differentiation in *C. Elegans* (Kaech et al. 1998). In this model, terminal differentiation of vulva epithelial cells from pluripotent precursors is dependent on the activation of basolateral membrane receptor tyrosine kinase, LET-23, by an interstitial EGF-like paracrine factor. Consequently, mutations in LET-23 or anything that disrupts the LET-23-dependent signal transduction cascade will produce a vulva-less phenotype.

Using a powerful combination of genetic, cell biological and biochemical analyses, Stuart Kim's group (Simske et al. 1996; Kaech et al. 1998) discovered that the null mutations in *lin-7, lin-2, or lin-10* yield a vulva-less phenotype in *C. Elegans* because the Let-23 receptor becomes mislocalized to the apical membrane, effectively separating the receptor from its ligand. As part of a multidisciplinary effort to understand how the three different PDZ proteins collectively determine the basolateral localization of LET-23, Kaech et al. tested for the possibility of biochemical association. Lin-7, Lin-10, and Lin-2 were found to form a heterotrimeric protein complex on the basolateral membrane. In vitro binding and yeast-two hybrid studies revealed that Lin-7 acts as the upstream scaffolding molecule, binding directly to the Let-23 receptor through a type 1 PDZ interaction (see Table 1) and with Lin-2 via a unique N-terminal interaction site, called a L27 domain (Simske et al. 1996; Kaech et al. 1998). To determine conclusively whether the interaction directly coordinates basolateral membrane localization of the receptor in vivo, Kaech et al.

implemented a series of clever experiments, exploiting the genetic tractability of *C. Elegans*. In these studies, the PDZ binding site in *LET-23* was altered in vivo to render the receptor unable to bind to the wild-type Lin-7 protein. A complementary mutation was introduced into the *lin-7* gene so that its protein product would only interact with the mutant LET-23 and not the wild-type receptor. As predicted, the mutant LET-23 receptor, lacking the type I PDZ binding motif, showed apical mislocalization in the wild-type Lin-7 background as did the wild-type receptor in transgenic animals carrying the mutant Lin-7. Even more convincingly, an allele-specific suppression of both the apical receptor mislocalization and the vulva-less phenotype was observed in double transgenic animals expressing the mutant receptor and the mutant Lin-7.

Exactly how Lin-7, Lin-2, and Lin-10 organize polarized expression of the receptor in VPC remains unknown. Nevertheless, an intriguing hypothesis that the PDZ complex might act in several different steps of a hierarchical targeting program (Borg et al. 1998; Butz et al. 1998) was put forward from work in other systems. Just as the work in *C. Elegans* was unfolding, orthologous gene products (Lin-7=Veli/MALS; Lin-2=CASK; Lin-10=Mint-1/X11) were identified as a tripartite protein complex in the mammalian brain (Butz et al. 1998), supporting the notion of an evolutionarily conserved mechanism for compartmentalizing proteins at specific membrane domains. Based on the known binding capacities of each component of the complex, it was suggested that the complex might act in two different steps of a membrane protein-targeting program in neurons (Borg et al. 1998; Butz et al. 1998). Interestingly, KIF17, a neuron-specific molecular motor in neuronal dendrites, interacts with the PDZ domain of mLin-10 (Mint1/X11), providing a mechanism for selective intracellular transport of vesicles containing the complex (Setou et al. 2000). The Lin-10 ortholog also interacts with a syntaxin binding protein called Munc18-1 (Hata et al. 1993). Borg et al. (1998), therefore, suggested that the tripartite complex might also link specific vesicle cargo to the SNARE machinery and provide a mechanism for docking exocytic vesicles to the appropriate membrane domain. Once docked, Lin-7, CASK and Lin-10 then would act as a retention complex, recruiting various membrane proteins into clusters at the membrane via different types of PDZ interactions. For instance, the PDZ domain in Lin-2/CASK binds the synaptic adhesion molecule neurexin (Butz et al. 1998), possibly anchoring the complex at the synaptic junction.

While this is an intriguing hypothesis for protein targeting in neurons, the biology of Lin-10 and Munc-18 in the kidney provides reason to suspect a significant divergence in mammalian epithelia. The mammalian counterpart of Lin-10 is actually encoded by a family of proteins called the Mints or X11s (Okamoto and Sudhof 1997). Although all three members of the Mint family share C-terminal PDZ and PTB domains, only the neuron-specific form, Mint-1, contains a CASK interaction domain (Borg et al. 1998). This observation suggests that the Mint module is dispensable for polarized targeting in the mammalian kidney in contrast to the absolute requirement for Lin-10 in the worm VPC (Kaech et al. 1998). The alternative view that an undefined Mint isoform assembles with CASK in the kidney cannot be ruled out. Since a Munc-18 isoform, Munc-18-2, has been implicated in apical, rather than basolateral membrane sorting (Riento et al. 1998), however, such a model might also necessitate a completely different mechanism than has been proposed for synaptic vesicle trafficking. Other Munc-18 isoforms have not, yet, been implicated in basolateral membrane trafficking. In any regard, the system loses the obvious link to microtubule-dependent delivery and Munc-18-dependent fusion, in the absence of the Mint-1 module.

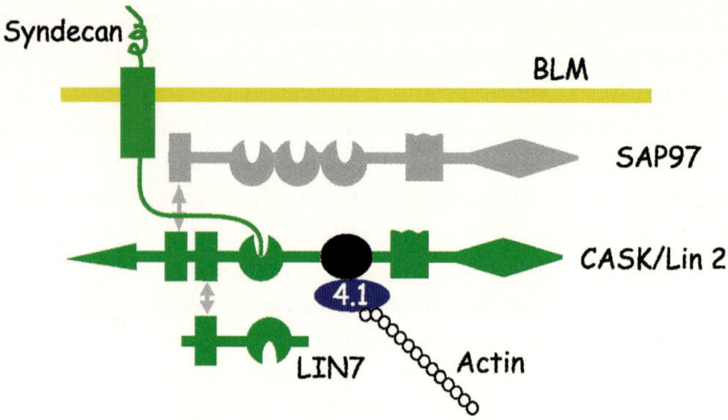

Fig. 3 The Lin 7/CASK/SAP 97 PDZ complex acts as basolateral membrane retention complex. CASK simultaneously interacts with two other PDZ proteins, Lin-7 and SAP-97, as well as with syndecan and the actin binding protein, 4.1, at the basolateral membrane. By linking extracellular matrix receptors and the cytoskeleton, the LIN7/CASK/SAP-97 complex has the capacity to act as a stable anchor to retain proteins that interact with the PDZ domains in Lin-7 and SAP-97 at the basolateral membrane. Box summarizes the functional domains present in members of the complex. L27 is protein–protein interaction domain first described in *Lin2/Lin7*, GUK is a guanylate kinase homology domain, SH3 is a Src homology 3 domain; the Hook domain interacts with 4.1 actin binding proteins, the CII kinase domain resembles the Ca2+/calmodulin-dependent protein kinase catalytic domain

Available evidence in mammalian epithelia cells supports the idea that Lin-7/CASK primarily functions to retain target proteins at the basolateral membrane. As predicted for a retention complex, CASK and Lin-7 coimmunoprecipitate and colocalize on the basolateral membrane of MDCK and native kidney epithelial cells (Straight et al. 2000; Olsen et al. 2002). Basolateral membrane localization of Lin-7 requires binding to one of the two L27 domains in CASK (Straight et al. 2000), while CASK associates with the basolateral membrane through a web of interactions. Indeed, multiple protein–protein interaction sites allow CASK to interact simultaneously with Lin-7, SAP97 (Lee et al. 2002), the heparin sulfate proteoglycan, syndecan, as well with the actin/spectrin binding protein, 4.1 (Cohen et al. 1998; Lee et al. 2002) (see Fig. 3). By linking extracellular matrix receptors and the cytoskeleton, the Lin-7/CASK complex has the capacity to act as a stable anchor to retain Lin-7 interacting proteins, like epithelial GABA transporter, BGT-1 (Perego et al. 1999), and the Kir 2.3 channel (Olsen et al. 2002), on the basolateral membrane. Indeed, Perego and colleagues found that removing the PDZ ligand in BGT-1 disrupted Lin-7 association in MCDK cells and dramatically increased the internalization of the transporter from the plasmalemma, consistent with a PDZ-dependent retention mechanism. Moreover, observations from heterologous expression studies that Lin-7 is sufficient to stabilize the Kir 2.3 channel on the cell surface provides direct evidence for Lin-7 as a retention factor.

Lin-7 appears to operate primarily as a component of a basolateral membrane retention machine in mammalian epithelial cells. Nevertheless, it should be pointed out that disruption of Lin-7 interactions can produce a wide range of mislocalization phenotypes, depending on the Lin-7 binding partners and the types of sorting signals embedded within them. For instance, mutant BGT transporters that lack their PDZ binding motif are predominately localized on the basolateral membrane. In this case, BGT-1 transporters are presumably directed to the basolateral membrane by non-PDZ-dependent sorting signals (Perego et al. 1997). Mutant Kir2.3 channels that lack their PDZ binding motif, on the other hand, are directed largely to an endosomal compartment rather than the basolateral membrane (Olsen et al. 2002). This is consistent with the presence of strong endosomal targeting signals that dominate when the retention motif is lost. In more marked contrast, removal of the PDZ binding site from a chimeric LET-23/ nerve growth factor receptor protein produces an apical-missorting phenotype (Straight et al. 2001). In this case, Lin-7 may normally stabilize the receptor on the basolateral membrane and thereby abrogate endosomal routing. In this scheme, a basolateral to apical transcytotic signal does not have the opportunity to be read in the postendocytotic pathway. The intriguing alternative possibility that Lin-7 acts more directly in the postendocytotic pathway to appropriately recycle the interaction partners back to the basolateral membrane remains to be rigorously explored.

Rafts

Membrane microdomains that are rich in glycosphingolipid and cholesterol are believed to exist as lipid "rafts" (reviewed in Brown and London 1998), which function as venues for specialized membrane trafficking and signal transduction operations (reviewed in Ikonen 2001). Initially, lipid rafts were hypothesized to operate as sorting platforms and delivery vehicles for apical membrane proteins and lipids traveling in the biosynthetic pathway. Simons and Van Meer (1988) proposed that lipid-raft carriers are vectorially delivered to the apical membrane after formation in the TGN. Consequently, selective assimilation of newly synthesized proteins into glycosphingolipid-rich microdomains of the TGN should guarantee targeting to the apical membrane. Since this appealing apical transport scheme was first proposed over a decade ago, its simple hypothesis has been challenged by exhaustive experimentation using a plethora of different apical membrane proteins and model systems. Evidence now indicates that lipid rafts can play supporting roles in apical trafficking via mechanisms that are far more complex than originally hypothesized.

Work with glycosylphosphatidylinositol (GPI)-anchored proteins illustrates the evolution of the raft hypothesis. Because GPI-anchored proteins accumulate in detergent-resistant membrane (DRM) compartments that are characteristic of lipid rafts (see Brown and London 1998 for a review on raft biochemistry), these proteins provide a convenient substrate to test a central tenet of the raft hypothesis, namely that apical sorting signals are decoded solely by lipid-raft partitioning.

In MDCK cells, early work indicated that GPI-anchored proteins, such as Placental Alkaline Phosphatase (PLAP), preferentially partition into a DRM compartment during biosynthetic transport and are subsequently targeted directly to the apical membrane (Brown and Rose 1992; Lisanti et al. 1990). With observations indicating that nearly all endogenous GPI-anchored proteins in MDCK cells are expressed on the apical membrane (Zurzolo et al. 1993), the correlation between GPI-linkage, DRM association and apical membrane delivery suggested that GPI anchors might serve as raft-dependent apical sorting signals. This view was initially reinforced by polarized trafficking studies of chimeric reporter constructs (Brown and Rose 1992; Brown et al. 1989), as well as GPI-anchored and transmembrane splice variants of a cell adhesion molecule, N-CAM (Powell et al. 1991). The results of these studies were reasonably interpreted to indicate that GPI-conjugation is necessary and sufficient to direct proteins to the apical membrane. However, the generality of this conclusion was convincingly challenged by subsequent studies. These studies revealed that polarized targeting of the chimeric reporter constructs could be largely explained by the presence of GPI-independent sorting signals (see Brown and London 1998 for review). Other examples were also identified. For instance, N-glycans can mediate apical sorting of some GPI-anchored, raft-associated proteins in MDCK cells (Benting et al. 1999). Additionally, in Fisher rat thyroid cells, where the majority of endogenous GPI-anchored proteins are paradoxically delivered to the basolateral membrane (Zurzolo et al. 1993), the prototypical GPI-anchored apical marker protein, PLAP, is expressed on the apical membrane by a GPI-independent mechanism (Lipardi et al. 2000). Furthermore, although disruption of lipid rafts in MDCK cells (by cholesterol depletion or through inhibition of glycosphingolipid synthesis) can cause missorting of some GPI-anchored proteins (Mays et al. 1995), others are not affected (Lipardi et al. 2000; Hannan and Edidin 1996). These results imply that raft-association is not always necessary for apical membrane trafficking, and although GPI anchors may certainly mediate raft association, they are often insufficient for apical sorting. Finally and surprisingly, recent studies indicate that GPI-linked proteins actually travel via an indirect pathway to the apical membrane in MDCK cells rather than a direct pathway as initially thought. Amazingly, they are delivered first to the basolateral membrane and then move to the apical surface by transcytosis (Polishchuk et al. 2004). Thus, if raft-dependent sorting is involved, the segregation process must occur at the plasmalemma or in the postendocytic pathway, rather than within the TGN as originally believed.

Transmembrane-domain dependent raft association has also been considered. The well-studied apical influenza proteins, hemagglutinin (HA) and neuraminidase, contain apical sorting determinants within their transmembrane domains (Lin et al. 1998; Kundu et al. 1996). The transmembrane domains in HA and neuraminidase also mediate association with DRM in the biosynthetic pathway. Disruption of DRM by cholesterol removal can cause missorting of these proteins to the basolateral membrane (Keller and Simons 1998, but see Tall et al. 2004). While depletion of cholesterol may have pleiotropic effects, these observations are nevertheless consistent with the idea that transmembrane domain interactions with DRM can coordinate apical membrane sorting processes. Further characterization of the apical sorting determinants in the transmembrane domains of HA (Lin et al. 1998; Tall et al. 2004) and neuramindase (Kundu et al. 1996) revealed that residues in the transmembrane domains that are required for apical expression do not necessarily correspond to the residues that are required for partitioning into DRM, however. Once again, rafts appear to be necessary but not sufficient for apical membrane targeting.

To complicate matters further, DRM association does not absolutely correlate with apical membrane targeting. Some basolateral membrane proteins also partition into raft-like compartments (Melkonian et al. 1995; Sargiacomo et al. 1993). For example, specialized cholesterol-enriched raft domains called caveolae (or caveolin protein-associated membrane compartments) are largely expressed on the basolateral rather than the apical membrane in MDCK cells (Mora et al. 1999; Lahtinen et al. 2003), intestinal cells (Vogel et al. 1998), and in the native kidney (Breton et al. 1998). Clearly, trafficking of lipid rafts in epithelial cells does not occur in the straightforward manner originally imagined. Since the raft theory requires unidirectional lipid trafficking, it is highly unlikely that apical sorting processes in the biosynthetic or indirect pathway are merely driven by a generic partitioning processes. Conceivably, extremely specific compartmentalization of proteins destined for the apical membrane in specialized (Roper et al. 2000) or noncaveolin II-containing glycosphingolipid-rich microdomains (Lipardi et al. 1998) could play a role.

Raft-associated trafficking molecules

Several observations described above raise the possibility of hierarchical apical sorting schemes that involve rafts. In these cases, raft association may simply be the price of admission to other necessary trafficking events. Segregation, delivery, or membrane-specific fusion machineries that are compartmentalized in rafts could sequentially process the particular apical membrane-destined proteins, which rely on raft-dependent sorting processes. Consistent with this idea, several raft-associated proteins have been identified as candidates for raft-dependent sorting machinery.

MAL. MAL/VIP17, a 17 kDa integral membrane protein (Alonso and Weissman 1987) containing a MARVEL domain (Sanchez-Pulido et al. 2002), is perhaps the best-characterized raft-associated sorting candidate. The unusual "lipid-like" properties of MAL (Millan et al. 1997), allow it to take up residence in special DRM, identified as apical membrane transport carriers in MDCK (Zacchetti et al. 1995) and Fisher thyroid epithelial cells (Martin-Belmonte et al. 2000). Antisense-mediated depletion of MAL in MDCK cells reduces HA inclusion into DRM and causes missorting of HA (Puertollano et al. 1999) and other DRM-associated apical membrane proteins (Cheong et al. 1999) to the basolateral membrane, supporting the role for MAL in a raft-dependent, apical-membrane trafficking process. Since the majority of HA interacts with MAL in DRM only after HA has reached the apical membrane, it is likely that MAL participates in trafficking steps following segregation and delivery of apical membrane proteins in the biosynthetic pathway (Tall et al. 2004). Consistent with this idea, roles for MAL in postendocytotic trafficking steps have recently been described (Martin-Belmonte et al. 2003). Furthermore, the missorting effects of MAL antisense on many different apical membrane proteins (Martin-Belmonte et al. 2000), including non-DRM associated proteins, implies that apical transport function(s) of MAL extend beyond raft-dependent processes.

Annexin. Members of the annexin family of calcium-dependent phospholipid-binding proteins, particularly annexin II (Jacob et al. 2004) and annexin 13b (Fiedler et al. 1995; Lafont et al. 1998) associate with DRM and are required for raft-dependent trafficking of some apical membrane proteins. Annexin 13b directly interacts with an apical membrane-directed microtubule motor protein, KIFC3, suggesting a mechanism that links DRM association with an apical membrane-directed delivery process (Noda et al. 2001). As an interesting possible corollary, annexin II associates with a population of post-Golgi vesicles that contain an actin motor protein, myosin 1a, which has been implicated in apical membrane delivery (Jacob et al. 2003).

Polarized delivery

For polarized protein expression to be achieved within biosynthetic or postendocytic recycling pathways, classical concepts suggest that the intracellular sorting operations, which segregate apically-directed molecules from basolaterally-directed ones, must be pursued by other trafficking processes. Most importantly, these include: (1) *direct delivery* of transport carriers to the appropriate acceptor membrane, or (2) *specific docking* and fusion with the correct membrane domain, or (3) some combination of both. A great deal of research has focused on determining whether the cytoskeleton and associated motor proteins carry vesicles from intracellular sorting compartments to polarized plasmalemma locales. Here, we provide a current understanding, highlighting recent developments in the field. The interested reader is also directed to two comprehensive reviews on the fundamental role of the cytoskeleton in epithelial polarization, motility, cell-to-cell adhesion, endocytosis, and vesicle delivery (Hamm-Alvarez and Sheetz 1998; Apodaca 2001).

Microtubule-mediated intracellular traffic

Microtubules (MT) are well known to facilitate the rapid (~0.1–1 μm/s) and efficient movement of vesicles to the plasmalemma. MT-dependent transport processes also can apparently facilitate selective plasmalemma targeting. The extent to which the MT tracking system is absolutely necessary for polarized targeting, however, appears to depend on the extent to which downstream processes, particularly docking and fusion, specify the final step in the polarized sorting program. Because the polarized localization of the docking and fusion machinery can depend on the MT system, it can be difficult to sort out the relative importance of the MT-dependent vesicle tracking mechanisms.

The unique organization of MT in polarized epithelial cells (Bacallao et al. 1989) along with directional requirements of MT-motor proteins (reviewed in Vale 2003) has long suggested a capacity for selective apical and basolateral membrane transport. Formed by polymerization of tubulin into protofilaments, MTs align in head-to-tail fashion to produce an intrinsically polarized plus- and minus-ended structure. In fully differentiated polarized epithelial cells, MT not only form a transverse network within the apical and basal cytoplasm, but also align into polarized bundles along the apical-to-basolateral axis (see Fig. 1). This contrasts with the broad radial organization that is commonly observed in fibroblasts and nonpolarized epithelial cells. Minus ends, anchored at a MT organizing center, face towards the apical membrane while the fast-growing plus ends extend toward the basolateral surface (Bacallao et al. 1989). Consequently, minus-end-directed motors, like dynein and the unconventional kinesins, are expected to selectively carry vesicles toward the apical membrane. Conventional kinesin motor proteins, on the other hand, are plus end directed. Consequently, they should transport membrane-destined vesicles to the basolateral membrane as well as track apical membrane vesicles along the mixed polarity subapical MT network.

A number of studies support this general model. Nevertheless, it should be pointed out that epithelial polarity can be established in the absence of the strategic longitudinal MT organization (Grindstaff et al. 1998a). These authors used classic pulse-chase, plasma membrane domain-selective detection techniques to assess biosynthetic processing of several apical and basolateral membrane proteins during the development cell surface polarity in the MDCK model. The result of these studies provided convincing evidence that direct targeting pathways from the TGN to the apical or basolateral membrane can become established before the longitudinal organization of MTs develops. While provocative, these observations should not be interpreted to indicate that MTs do not normally participate in some fashion, however minor, to polarized trafficking. If the MT system acts within an intermediate step, its contribution to the polarized sorting processes can be difficult, if not impossible, to detect with classical techniques, particularly when surface membrane-specific docking and fusion processes are in play.

With the development and application of sophisticated real-time imaging techniques (Kreitzer et al. 2003), it has now become possible to directly monitor the intracellular trafficking of post-Golgi transport vesicles. Using these techniques, Kreitzer et al. assessed the role of MTs in trafficking and fusion of post-Golgi transport intermediates (PGTI), carrying GFP-tagged apical or basolateral-destined proteins. Before polarity develops in MDCK cells, apical and basolateral PGTI randomly distribute throughout the cytoplasm and fuse randomly with the basal domain of plasmalemma. After polarization (see section entitled "Genesis of polarity," below) and coincident with the development of the axial organization of microtubules, PGTI trafficking and fusion appears to become more directed and plasma membrane domain-specific. MT disrupting agents caused apically destined PGTI to fuse with the lateral membrane. Interestingly, the appropriate localization of the apical docking protein, syntaxin 3, (see section entitled "Docking and fusion," below) was also dependent on MTs; MT-disrupting agents caused syntaxin 3 to become randomly distributed to apical and basolateral membrane domains. The result extends the role of MTs beyond just a route for polarized delivery of vesicles containing apically destined cargo. In contrast, MT-disrupting agents did not inhibit delivery or fusion of transport vesicles containing basolateral cargo (LDLR or NCAM) to the basolateral membrane.

These observations are consistent with numerous studies, which show that efficient transport to the apical plasmalemma is highly sensitive to MT disrupting agents such as nocodazole or cholchicine. Indeed, drug-mediated MT dispersion significantly slows delivery of membrane proteins to the apical membrane whether they are traveling in the biosynthetic pathway (Breitfeld et al. 1990; Grindstaff et al. 1998a), the apical recycling pathway (Apodaca et al. 1994), or the transcytotic pathway (Maples et al. 1997). For some apical proteins, MT depolymerization causes missorting to the basolateral membrane (Sabolic et al. 2002; Saunders and Limbird 1997; Rindler et al. 1987). Such studies suggest that apical membrane-directed MT-dependent traffic can also specify the final polarized sorting operation of certain select proteins. Without it, promiscuous docking and fusion reactions prevail.

As predicted by the dependence on an intact MT cytoskeleton, efficient transport to the apical membrane requires MT motors. LaFont et al. showed that immunodepletion of dynein or kinesin inhibits the transport of HA, the prototypical apical marker, to the apical membrane of MDCK cells (Lafont et al. 1994). Using time-lapsed imaging to study intracellular trafficking of a GFP-tagged apical marker, p75, Kreitzer et al. not only confirmed that kinesin and dynein are required for post-Golgi transport, but also provided evidence that different motor proteins may play distinct roles (Kreitzer et al. 2000). They found that dissimilar mistrafficking phenotypes are produced by immunodepletion of unconventional kinesin and expression of dominant negative dynein. Consequently, they suggested that kinesin is essential for the formation and transport of apically-directed tubulo-vesicular carriers while dynein is required for release of vesicles from the TGN. Consistent with these observations, KIFC3, a minus end-directed kinesin that is associated with subapical transport vesicles in kidney epithelial cells, has been shown to drive transport of vesicles containing HA and annexin XIIIb (see section entitled "Rafts") to the apical membrane (Noda et al. 2001). The role of dynein extends beyond the TGN, however. For instance, in the renal collecting duct, dynein associates with a population of post-TGN intracellular vesicles, containing AQP2 water channels (Marples et al. 1998). These aquaporins are translocated to the apical membrane in response to vasopressin-stimulation through an MT- (Sabolic et al. 1995) and dynein- (Marples et al. 1996) dependent mechanism. The interaction between AQP2 and dynein is most likely indirect, however, and is likely to be governed only by common affinity to the same population of vesicles. Nevertheless, it should be pointed out that recent studies provide a precedent for a more direct mechanism. Importantly, TGN-to-apical membrane trafficking of rhodopsin is coordinated by a direct interaction between its apical sorting signal (Chuang and Sung 1998) and the dynein light chain, Tctex-1 (Tai et al. 2001).

Basolateral trafficking is more resistant to MT disruption (Boll et al. 1991; Breitfeld et al. 1990). For instance, some secreted proteins are dependent on MTs for polarized delivery to the basolateral membrane while others are not (Boll et al. 1991). LaFont et al. reported that either nocodazole or kinesin immuno-depletion can inhibit basolateral transport of VSV-G in permeabilized MDCK cells (Lafont et al. 1994). More recent studies indicate that complete MT depolymerization in intact MDCK cells only reduces the rate of delivery of newly synthesized membrane proteins to the basolateral membrane without causing profound missorting to the apical membrane (Grindstaff et al. 1998a). Thus, it is likely that many basolateral-destined vesicles travel by alternative, MT-independent transport pathways.

Actin-mediated intracellular traffic

Actin filaments, like MTs, have intrinsic fast-growing plus ends ("barbed") and slower-growing minus ends ("pointed"). Unlike MTs, however, actin filaments do not organize into apical to basolateral-directed tracks, but rather arrange themselves into dense networks or into loose nonparallel bundles known as stress fibers. Actin filaments also tend to be shorter than MTs and to conduct cargo more slowly. Transport along their networks is driven by myosin motor proteins (Mermall et al. 1998). All myosins move toward the barbed (+) end of actin filaments except for myosin VI (Wells et al. 1999).

Investigations into the role of actin in polarized trafficking operations have generally relied on drugs that disrupt actin filament polymerization dynamics. Unfortunately, these studies have produced conflicting and confusing results. Treatment of MDCK cells with cytochalasin D (CD), a fungal toxin that caps the barbed ends of actin filaments, impairs endocytosis from the apical membrane to the AEE (Gottlieb et al. 1993). This inhibits apical membrane-directed transcytosis and alters apical recycling (Maples et al. 1997) without affecting endocytosis or recycling of several basolateral marker proteins. In contrast, treatment of MDCK cells with latrunculin B (LatB), another actin-depolymerizing agent, was shown to specifically alter postendocytotic trafficking of basolateral but not apical cargo proteins (Sheff et al. 2002). Jasplakinolide, a drug that stabilizes actin filaments, increases uptake of fluid-phase markers from the basolateral, but not the apical surface (Shurety et al. 1998). While these contradictory reports are difficult to reconcile, the positive effects of different drugs suggest that the actin cytoskeleton may play roles in intracellular trafficking to both apical and basolateral membranes. We anticipate that future studies, using molecular rather than pharmacological-based techniques, may provide a clearer picture of actin-dependent intracellular vesicle trafficking within epithelial cells.

Some progress has been made in identifying the myosin motors responsible for actin-based trafficking in polarized epithelial cells. Work on Class V myosins, implicated as motors for actin-dependent organelle and vesicle transport (reviewed by Langford 2002), is particularly intriguing. Myosin Vb directly interacts with Rab 11a in common recycling endosomes (Hales et al. 2002). As judged by the selective effects of negative-interfering constructs on polarized postendocytotic recycling, mysosin Vb appears to function in the apical recycling system (Lapierre et al. 2001). Another Class V myosin isoform expressed primarily in epithelial cells, myosin-Vc, colocalizes with and functions in a Rab 8-dependent (see section entitled "Rab") endocytic compartment (Rodriguez and Cheney 2002). These tantalizing preliminary results suggest a schema whereby myosin-Vb mediates apical recycling, while myosin-Vc assists with basolateral traffic. Obviously, more data are required to verify these roles.

The backward-moving myosin, myosin VI, has recently been localized within an intermicrovillar coated-pit region of the apical brush border membrane in the kidney proximal tubule (Biemesderfer et al. 2002). Myosin VI has the capacity to associate with clathrin-coated vesicles and membrane protein cargo at the plasma membrane. Indeed, the cargo-binding tail domain of myosin VI interacts with scaffolding and adaptor molecules, such as Dab2 (Morris et al. 2004), GIPC (Simon et al. 1998), and SAP 97 (Bunn et al. 1999). The myosin VI may function to move exocytic vesicles towards the minus end of actin filaments, away from the plasma membrane, and into the cell.

A candidate motor protein has also been identified as the mediator for the actin-dependent release of vesicles from the TGN. Musch et al. demonstrated that the TGN coat protein, p200, contains a myosin II domain that is critical in the assembly of VSVG-containing vesicles in MDCK cells. Depletion of p200/myosin results in a 50% reduction in VSVG vesicle release from the TGN (Musch et al.1997). Others have found that removal of this protein from the cytosol does not affect vesicle generation; instead the p200/myosin antibody used by Musch could interact with undenatured coatomer, a protein that they found essential for vesicle formation (Simon et al. 1998). In light of this, the identity of the actin motor that assists with vesicle release remains in question.

Polarized docking and fusion

Specific membrane fusion machines are thought to provide a critical and necessary level of control in a hierarchical scheme of polarized membrane protein targeting, guaranteeing that vesicles fuse with the appropriate target membranes while preventing misguided vesicles from accidentally uniting with the wrong acceptor membrane.

SNARES

The core machinery for vesicle fusion (Sollner et al. 1993b; Rothman 1994; Jahn and Sudhof 1999; Chen and Scheller 2001) is comprised of a family of structurally related integral membrane proteins, called the SNAREs (*s*oluble *N*-ethyl maleimide-sensitive *a*ttachment factor *re*ceptors), and two soluble proteins called NSF (*N*-ethyl maleimide-sensitive factor) and α-SNAP (soluble NSF attachment factor). Interaction between specific SNAREs on transport vesicles (v-SNAREs, generally VAMP/synaptobrevin homologs) and their cognate SNAREs on target membranes (t-SNAREs, generally homologs of syntaxin and/or SNAP-25) is thought to form stable, membrane-specific docking complexes (Fasshauer et al. 1998; Sutton et al. 1998). Because the intertwined v-/t-SNARE heterotrimeric complex has an unusually high melting point (70–90°C), its formation is believed to provide the free energy for membrane fusion (Sutton et al. 1998; Jahn and Sudhof 1999). Disassembly of the SNARE complex is then controlled by the direct interaction with α-SNAP in association with the ATP bound form of NSF. SNAP stimulates ATPase activity on NSF to trigger SNARE disassembly (Sollner et al. 1993a; Barnard et al. 1997). According to the original SNARE hypothesis (Rothman 1994), fusion selectivity is achieved by the specificity of pairing between v-SNAREs and t-SNAREs. Although this innovative proposal still remains somewhat controversial and may now require some refinements (Chen and Scheller 2001; Waters and Pfeffer 1999; Guo et al. 2000), it triggered a flurry of productive research activity in membrane biology, including the polarized trafficking field.

Early studies suggested that apical membrane-directed traffic in the biosynthetic pathway might involve a SNARE-independent mechanism (Ikonen et al. 1995; Fiedler et al. 1995; Verkade and Simons 1997). Subsequent work has proven that the apical membrane protein trafficking does not, in fact, depend on mechanisms that are profoundly different from other known membrane fusion processes. SNARE machinery is certainly required for apical trafficking of marker proteins in MDCK cells (Apodaca et al. 1996; Low et al. 1998a; Lafont et al. 1999) and physiologically relevant transport proteins, such as the AQP2 water channel (Jo et al. 1995; Nielsen et al. 1995; Inoue et al. 1998; Mandon et al. 1997; Gouraud et al. 2001) and proton ATPases (Frank et al. 2002; Banerjee et al. 1999, 2001) in the native kidney, bladder (Franki et al. 1995), and male reproductive tract (Breton et al. 2000).

Localization studies have, in fact, suggested a capacity for specific, t-SNARE-dependent fusion processes that are individually tailored for either apical or basolateral membrane domains. Indeed, some t-SNARES display a remarkable polarized distribution in epithelia, quite unlike the nonspecific plasmalemma compartmentalization in neuronal cells (Foletti et al. 1999). Studies on SNAREs that are endogenous to MDCK cells provided early and noteworthy examples. In contrast to the nonpolarized distribution of SNAP-23, a ubiquitous SNAP-25 homolog, and syntaxins 2A and 11, Low reported that Syntaxin 3 and Syntaxin 4 are expressed on opposite membrane domains of MDCK cells. Based on these observations they speculated that these particular t-SNARES might underpin polarized vesicle fusion processes (Low et al. 1998b, 1996, 2000). Several lines of evidence are consistent with this idea. First, over expression of Syntaxin 3, the apically-localized t-SNARE in MDCK cells, selectively inhibits apical membrane traffic (Low et al. 1998a). Second, Lafont et al. reported that addition of an anitsyntaxin 3 antibody to streptolysin-O-permeabilized MDCK cells selectively decreases reporter membrane protein traffic to the apical membrane. Addition of an anitsyntaxin 4 antibody, in contrast, selectively decreases membrane protein traffic to the basolateral membrane (Lafont et al. 1999).

The polarized distribution of endogenous Syntaxin 3 and Syntaxin 4 in the native epithelial cells of the kidney is controversial, however. Several groups have reported that Syntaxin 3 is found on the basolateral membrane while Syntaxin 4 is expressed on the apical membrane (Breton et al. 2002; Mandon et al. 1996, 1997), precisely the opposite of what is observed in MDCK cells (Low et al. 1996), Ca-Co 2 cells, and the native intestine (Delgrossi et al. 1997). Li et al., however, have recently reported that syntaxin 3 exclusively localizes to the apical membrane and syntaxin 4 localizes to the basolateral membrane of all epithelial cells in the kidney (Li et al. 2002). At present the explanation for the contradictory reports of syntaxin 3 and 4 localization is unknown. Several alternative-splice forms of syntaxin 3 have been identified (Ibaraki et al. 1995), raising the possibility of isoform-specific, differential localization as has been shown for Syntaxin 2A and B (Quinones et al. 1999).

Three different vesicle-associated SNAREs have also been implicated in polarized membrane traffic, at least at the apical membrane. In the renal cortical collecting duct principal cell, VAMP2 (Rossetto et al. 1996), along with SNAP-23, is present in subapical intracellular vesicles that contain the AQP-2 water channel. This is consistent with a role in vasopressin-regulated trafficking at the apical membrane (Inoue et al. 1998; Nielsen et al. 1995). Likewise, VAMP2 associates with SNAP-23 in subapical vesicles containing the H^+-ATPase in intercalated cells (Banerjee et al. 2001). Another VAMP-isoform, VAMP 7 (also called TI-VAMP, for *T*etanus neurotoxin *I*nsensitive), forms complexes with SNAP-23 and syntaxin 3 at the apical membrane of CaCo-2 cells (Galli et al. 1998) and MDCK cells (Lafont et al. 1999). Furthermore, antibodies directed against VAMP7 inhibit HA traffic to the apical membrane. Steegmaier et al. used GFP-tagged v-SNAREs to monitor the trafficking between intracellular compartments and the polarized plasma membrane domains in MDCK cells (Steegmaier et al. 2000). They reported that VAMP8/endobrevin uniquely cycles through the apical but not through the basolateral membrane. Interestingly, VAMP8, an abundant v-SNARE in epithelia, is restricted to the apical membrane of nearly all segments of the nephron, suggesting that this v-SNARE may play a general role in apical endosomal trafficking in polarized epithelial cells. Additional studies are required to determine whether any v-SNAREs are uniquely associated with basolateral membrane traffic.

While the polarized distribution of some SNAREs in epithelia is intriguing, it should be pointed out that it remains to be determined whether fusion specificity is encoded solely by t-SNARE/v-SNAREs partnerships. Certainly, SNARE pairing promiscuity, among other factors, has made this issue the subject of tremendous debate (see Pelham 2001 for recent update) in membrane biology. In the special case of epithelial cells, the precise distribution of various SNAREs presently appears to be somewhat cell specific. Thus, it seems unlikely that any particular SNARE combination could provide a specific and universal apical or basolateral fusion role. SNARE-associated proteins, such as Munc-18 (Riento et al. 1998; Rowe et al. 2001) and the mammalian form of *Drosophila* tumor suppressor protein lethal (2) giant larvae (l(2)gl) (Musch et al. 2002), may have roles in specifying polarized fusion. Recent studies in the general fusion arena have strongly suggested a role for a variety of other factors, such as membrane tethers and Rab proteins. These molecules appear to coordinate initial docking events preceding SNARE-mediated fusion (Waters and Pfeffer 1999). While significance of all these factors in polarized fusion events in epithelia remains to be determined, the exocyst might serve as a prototype.

Exocyst

The exocyst was first identified in genetic and biochemical studies in budding yeast (Terbush et al. 1996). It is a multiprotein complex which functions as a necessary component of the intracellular exocytic machinery and mediates polarized tethering and/or docking of post-Golgi vesicles at the growing bud tip. The complex, composed of eight proteins (sec3p, sec5p, sec6p, sec8p, sec10p, sec15p, Exo70p, and Exo84p), is evolutionarily conserved (Hsu et al. 1996; Matern et al. 2001). Present evidence strongly suggests that the exocyst governs a polarized trafficking operation in mammalian epithelia, targeting certain Golgi-derived or postexocytic vesicles to the basolateral membrane (Grindstaff et al. 1998b; Yeaman et al. 2001; Moskalenko et al. 2002; Matern et al. 2001). Perhaps, as a consequence, the exocyst is also required for normal renal tubulogenesis (Lipschutz et al. 2000; O'Brien et al. 2001; Charron et al. 2000).

Grindstaff and coworkers (1998) and Yeaman and coworkers (1999, 2001) first postulated that the exocyst operates as a selective docking site for the insertion of vesicles destined for the basolateral membrane. In MDCK cells, components of the exocyst, namely Sec 6/8 [because early work was largely performed with antibodies to Sec 6 and Sec 8, the mammalian exocyst is often called the Sec 6/8 complex (Hsu et al. 1999)], are rapidly recruited from the cytosol to sites of cell–cell contact upon initiation of calcium-dependent cell–cell adhesion (see section entitled "Genesis of cell polarity," below). As polarity develops, the complex becomes restricted to a zone near the junctional complex on the lateral membrane (Grindstaff et al. 1998b). In addition, functional studies demonstrated that neutralizing antibodies to Sec 8 partially inhibit the delivery of LDL receptors to the basolateral membrane but have no effect on apical membrane-destined cargo in the streptolysin-O permeabilized polarized MDCK cell model (Grindstaff et al. 1998b). Taken together, these observations provided key evidence for the involvement of the Sec 6/8 complex as an active site of insertion for basolateral membrane-destined transport vesicles. Recent live-cell imaging analysis of post-Golgi trafficking vesicles in MDCK cells indicated that plasma membrane fusion events of vesicles containing GFP-tagged LDLR, may in fact, occur along the lateral membrane at sites that are nearly, albeit not perfectly, coincident with Sec 6 localization (and syntaxin 4; see section entitled "Fusion") (Kreitzer et al. 2003). This new observation reinforces the concept that the Sec 6/8 complex may be involved in basolateral membrane docking.

Other components of the exocyst have been implicated in polarized basolateral membrane trafficking. For instance, Sec 10 colocalizes with Sec 6/8 (Lipschutz et al. 2000) and Exo 70 is recruited into the Sec 6/8 complex at cell–cell contact points in polarizing MDCK cells (Matern et al. 2001). Interestingly, overexpression of exogenous Sec 10 selectively enhances basolateral transport in MDCK cells (Lipschutz et al. 2000).

Our present understanding about how the exocyst might orchestrate polarized delivery, tethering, and/or docking of exocytic vesicles at the lateral membrane in mammalian epithelial cells has been shaped from mechanistic studies on the yeast exocyst. In yeast, assembly of the exocyst complex is tightly controlled by a specific secretory vesicle-associated Rab GTPase, called Sec 4 (see section entitled "Rab") (Guo et al. 1999). Docking, on the other hand, is specified by Rho-dependent localization of Sec 3 at the bud tip (Finger et al. 1998). In this scheme, the active, GTP-bound, form of Sec 4 directly interacts with a component of the exocyst, Sec 15. This protein–protein interaction then initiates assembly of the entire exocyst complex, and consequently Sec 4p on the secretory vesicle surface becomes indirectly tethered to Sec 3 on the cell surface (Guo et al. 1999). Intriguingly, the most likely mammalian ortholog of Sec 4, Rab 8, has been strongly implicated in basolateral membrane trafficking (Huber et al. 1993; Ang et al. 2003) (see sections entitled "Rab" and "Adaptin"), suggesting a parallel between yeast and mammalian epithelia. Further studies are required to determine if Rab 8 actually interacts with mammalian Sec 15p to orchestrate exocyst assembly.

Although there may be a number of mechanistic parallels, the mammalian exocyst in epithelial cells appears to be different from yeast in several important respects. First, intracellular localization of the Sec 6/8 complex may be highly dynamic. For instance, maneuvers that block trafficking in the secretory pathway, such as low temperature or Brefeldin A treatment, cause Sec 6/8 to accumulate near the TGN (Yeaman et al. 2001). These observations suggest that localization of Sec 6/8 in mammalian epithelia depends on continuous exocytic vesicle trafficking. Second, in contrast to yeast, the mammalian Sec 3p, is neither a Rho effector nor does it act as a spatial landmark for exocytosis on the lateral membrane of mammalian epithelia (Matern et al. 2001). Exactly how the complex associates with the lateral membrane remains unknown. Finally, the mammalian exocyst is regulated differently. Unlike yeast, the mammalian exocyst is a target of the Ral GTPase (Moskalenko et al. 2002, 2003; Sugihara et al. 2002). In mammalian systems, direct interaction of Ral GTPases with specific components of the exocyst, namely Sec 5p (Moskalenko et al. 2002; Sugihara et al. 2002) and Exo 84p (Moskalenko et al. 2003), can facilitate the assembly of the entire octameric exocyst complex (Moskalenko et al. 2002, 2003). As a consequence, perturbation of Ral signaling causes inhibition of exocyst assembly and missorting of several basolateral membrane proteins to the apical surface of polarized epithelial cells (Moskalenko et al. 2002, 2003).

In summary, the exocyst plays an important role in polarized targeting of basolateral membrane-destined post-Golgi vesicles. Additional studies are now needed to elucidate how and where specific cargos are selected for exocyst-dependent trafficking.

Genesis of cell polarity

It is generally thought that de novo generation of membrane polarity in epithelial cells is initiated by asymmetric spatial cues at the cell surface that involve cell–cell and cell–extracellular matrix (ECM) adhesion (reviewed in Rodriguez-Boulan and Nelson 1989). This key concept has its roots in the study of epithelial structure formation during embryogenesis. The development of a polarized epithelial cell phenotype in the kidney, following inductive interactions between the uretic bud and mesenchymal cells, correlates with the expression of adhesion proteins that mediate cell-to-cell and cell-to-substratum attachment (Ekblom et al. 1986; Klein et al. 1988). In the past decade, the notion has been buttressed by a growing body of experimental data generated with epithelial cells in culture. These studies indicate that epithelial polarization is indeed induced by specific cell–cell adhesion molecules, namely E-cadherin, a member of the Ca^{++}-dependent cadherin superfamily (Huber et al. 1996), and nectin, an immunoglobin-like adhesion molecule, as well as specific integrins and laminin, which coordinate cell adhesion to ECM. Cellular adhesion contacts not only serve as molecular landmarks to define contacting and noncontacting sites on the plasma membrane as forerunners of basolateral and apical membrane (Yeaman et al. 1999), but also trigger intracellular signaling cascades (Nelson and Nusse 2004). These, in turn, stimulate the local assembly of polarity protein complexes involved in tight junction formation and cytoskeletal rearrangements that are critical for development of the polarized phenotype.

The first hint that cell-to-cell adhesion might be involved in the epithelial polarization process came from observations that MDCK cells display a random distribution of membrane proteins when grown at a sufficiently low density to prevent cell–cell contacts. These cells then gradually become polarized as cellular confluence is reached, correlating with cell–cell contact development (Balcarova-Stander et al. 1984; Herzlinger and Ojakian 1984). Similar observations were made when MDCK cells were grown in the absence of extracellular calcium to prevent cadherin dependent cell–cell contacts (Ojakian and Schwimmer 1988; Vega-Salas et al. 1987a). Under these conditions, MDCK cells not only fail to polarize but also develop an unusual intracellular storage compartment. This structure, called a vacuolar apical compartment (VAC), contains apical membrane markers and microvilli (Vega-Salas et al. 1987b). Upon elevation of extracellular calcium, often referred to as a calcium-switch protocol (reviewed in Cereijido et al. 2000), cadherin-dependent contacts are formed and polarization ensues (Wang et al. 1990a; Rodriguez-Boulan et al. 1983). This occurs as polarity protein complexes (see section entitled "Junction development" below) and components of the tight junction, such as ZO-1, are recruited to a zone immediately adjacent to cell–cell adhesion points (Wang et al. 1990a; Vega-Salas et al. 1988; Rajasekaran et al. 1996). At the same time, apical and basolateral domain-specific submembrane cytoskeleton components begin to assemble (Salas et al. 1988; Nelson et al. 1990) and VAC exocytosis occurs towards the apical aspect of the forming tight junction (Vega-Salas et al. 1988).

Exogenous expression of E-cadherin in nonpolarized fibroblasts is sufficient to induce selective redistribution of Na/K ATPase to the E-cadherin-mediated cell–cell contact zones, comparable to the lateral distribution of Na/K ATPase in polarized epithelial cells (McNeill et al. 1990). Interestingly, this response occurs in the absence of ZO-1 and does not require the formation of a barrier junction. Instead, the process coincides with a reorganization of the actin cytoskeleton (Adams et al. 1996, 1998) and a recruitment of fodrin to sites of E-cadherin mediated cell–cell contact (Adams et al. 1996, 1998). Because fodrin and ankyrin interact with the alpha subunit of the Na/K ATPase (Nelson and Veshnock 1987; Koob et al. 1988; Morrow et al. 1989), it is generally assumed that the local organization of the membrane cytoskeleton at E-cadherin junctions effectively anchors the transport protein at the regions of cell–cell contact. Consistent with this idea, the cytoplasmic tail of E-cadherin, indispensable for the induction of cell surface redistribution of the Na/K ATPase, interacts with alpha and beta catenin to form stable linkages with the actin cytoskeleton and reinforce cell–cell adhesion (Jou et al. 1995; Adams et al. 1996; Nathke et al. 1994; Nelson et al. 1990). In MDCK cells, E-cadherin, alpha-catenin, and beta-catenin form complexes (Nathke et al. 1994) containing ankyrin and fodrin (Nelson et al. 1990), which localize along cell–cell contacts and associate with bundles of actin (Adams et al. 1996) and the Na/K ATPase.

The extent to which other cadherin isoforms in the kidney participate in cell–cell adhesion and polarization is not known. However, it is notable that cadherin-6, a type II cadherin, is expressed during early stages of nephrogenesis (Cho et al. 1998). Interestingly, targeted disruption of the cadherin-6 gene in mice delays the timing of the renal epithelial polarization process in the embryonic kidney (Mah et al. 2000). Like E-cadherin, cadherin-6, assembles in complexes with alpha- and beta-catenin at sites of cell–cell adhesion in MDCK cells. Unlike E-cadherin, cadherin-6 is negatively regulated by a Wnt/Wingless pathway (reviewed in Barth et al. 1997), suggesting additional roles in signal transduction (Stewart et al. 2000).

An entirely different cell–cell adhesion molecule, nectin, has also been implicated in epithelial polarization. Nectin forms calcium-independent adhesions through homophilic interactions involving extracellular immunoglobin-like domains (Yasumi et al. 2003). The intracellular COOH-terminus of nectin associates with filamentous actin through a PDZ adaptor protein, afadin (Asakura et al. 1999). Protein–protein interaction studies, combined with observations in heterologous expression systems, indicate that nectin has the capacity to recruit the E-cadherin-catenin complex (Yokoyama et al. 2001) through cytoplasmic tail interactions with afadin and catenin (Tachibana et al. 2000). Based on observations that dominant negative nectin interferes with E-cadherin complex formation and cell adhesion, it seems likely that these interactions are necessary to facilitate proper assembly of the cell–cell adherens junction (Honda et al. 2003). Nectin also associates with the tight junction proteins, ZO-1 (Yokoyama et al. 2001) and junctional adhesion molecule (JAM) (Fukuhara et al. 2002) as well as a key polarity protein, PAR-3 (see section entitled "Junction formation," below). This raises the intriguing possibility that nectin-based adhesion also plays a role in the formation of the tight junction. Perhaps it serves as a platform where signals, induced by cell–cell contact, are translated into tight junction assembly.

Although cell–cell contacts are necessary for the induction of epithelial polarity, they are not sufficient. Cell–extracellular matrix (ECM) interactions are also required. The development of MDCK suspension culture system (McAteer et al. 1987; Grantham et al. 1989), allowed experimental control over cell–substratum contact and provided an ideal model to test the role of ECM in polarization . Using this system, Wang et el. discovered that cell–cell contact is sufficient to trigger segregation of apical and basolateral marker proteins and to stimulate ZO-1 recruitment to the cell–cell contact zone, but cell–substratum contact is essential to induce the appropriate localization of tight junction proteins and the proper orientation of the apico-basolateral membrane axis (Wang et al. 1990a). The key concept that cell–ECM interactions define where the apical and basolateral membrane lie relative to the external environment is elegantly demonstrated by the pleomorphic behavior of MDCK cells grown in suspension culture. When seeded in fluid suspension, MDCK cells form distinctive multicellular cysts with an opposite polarity to cysts that are formed in a collagen (type IV) gel. In fluid suspension culture, the basolateral membrane faces a central lumen and the apical membrane faces the outside, whereas in collagen gel, the basolateral membrane faces the outside and the apical membrane faces the central lumen. Moreover, fully developed cysts that are grown in fluid suspension rapidly reverse cell polarity when placed on collagen gel substrate (Wang et al. 1990b). Upon apical membrane-outside contact with the collagen matrix, the apical membrane is rapidly internalized and degraded as a new apical surface forms on the free surface. Certain ECM molecules, particularly $\alpha2$ beta 1 integrins (Schwimmer and Ojakian 1995; Saelman et al. 1995) and laminin (O'Brien et al. 2001), have been identified as key adhesion receptor proteins that mediate the cell–ECM-induced axis of polarity.

Other ECM proteins also may play roles in cell-specific polarization processes. For instance, hensin, a 230-kDa ECM protein (van Adelsberg et al. 1994), appears to trigger polarized remodeling of intercalated cells in the kidney (reviewed in Al-Awqati 2003). Recent studies indicate that the bicarbonate secreting β-intercalated cells deposit hensin into the extracellular matrix in response to acidification. This stimulates the adaptive polarity conversion of β-intercalated cells into acid-secreting α-intercalated cells (Schwartz et al. 2002).

Junction development

Proper formation of the tight junction is a key underpinning of the polarization process in vertebrate epithelial cells. By acting as a molecular fence or diffusion barrier, this structure separates and demarcates the boundary between the apical and basolateral membranes and prevents intermixing of polarized membrane proteins (reviewed in Cereijido et al. 1998). Recent genetic studies in *Drosophila melanogaster* and *Caenorhabditis elegans*, complemented by biochemical and cell biological studies in mammalian epithelial cell culture, have begun to offer mechanistic insights into the formation of epithelial cell–cell junctions. While the organization of the junctions is quite different in vertebrates, worms, and the fly (reviewed in Knust and Bossinger 2002), the results of these studies indicate that the molecules that control their development are very similar in very divergent organisms. To date, three groups of orthologous gene products have been identified. Each group contains and is distinguished by the presence of one or more PDZ domain proteins. Perhaps best known by their names in the fly, the groups are called Stardust/Crumbs/Discs Lost; Bazooka/Par 6/aPKC; and Scribble/ Discs Large/ Lethal Giant Larvae.

Stardust (Sdt) is a membrane-associated guanylate kinase (or MAGUK, see section entitled "PDZ proteins"). It associates with Crumbs (Crb), an apically-oriented transmembrane protein, and Discs Lost (Dsl), a PDZ-containing protein, within a subapical domain in fly epithelia near the apex of the lateral membrane (Muller and Wieschaus 1996; Grawe et al. 1996). The site of assembly corresponds to the region where the tight junction develops in vertebrates. The PDZ domain of Stardust interacts with the cytoplasmic C-terminal tail of Crumbs (Tepass et al. 1990; Tepass and Knust 1993; Knust et al. 1993; Grawe et al. 1996; Bachmann et al. 2001), while its L27 binding domain associates with the MAGUK recruitment domain of Discs Lost (Roh et al. 2002a, b; Bhat et al. 1999), to form a ternary complex (see Fig. 4). In *Drosophila*, null mutations in Crumbs, Stardust (Muller and Wieschaus 1996; Tepass and Knust 1993; Knust et al. 1993), or Discs Lost (Bhat et al. 1999) cause defects in junction formation and alterations in apico-basolateral polarity. In contrast, overexpression of Crumbs leads to polarity defects and expansion of the apical surface (Wodarz et al. 1995). Such observations indicate that the Stardust/Crumbs/Discs Lost complex acts to coordinate the formation of apical membrane as well as the adherens junction in the fly.

The mammalian orthologs of Stardust (PALS1; Kamberov et al. 2000), Discs lost (Pals1-associated tight junction protein; PATJ), and Crumbs (CRB3; Makarova et al. 2003) participate in identical interactions, forming a ternary complex near the tight junction of renal epithelial cells (Roh et al. 2002b). While the many details about the function of PALS1-PATJ-CRB3 in mammalian epithelia remain poorly understood, early reports suggest that the complex participates in tight junction assembly. This is reminiscent of the role of Stardust-Discs Lost-Crumbs in adherens junction assembly in *Drosophila*. For instance, PATJ directly binds to tight junction proteins, ZO3, and claudin 1 through PDZ interactions and presumably links the complex to the tight junction. More importantly, expression of dominant negative PATJ in MDCK cells causes mislocalization of PALS and fragmented assembly of the tight junction (Hurd et al. 2003) but has no effect on the integrity of preformed junctions (Roh et al. 2002b). Interestingly, mutations in a human Crumbs isoform that is exclusively expressed in the eye, CRB1, have been linked with two severe forms of retinal dystrophy, Retinitis Pigmentosa 12 and Leber's congenital amaurosis (Den Hollander et al. 1999), raising the possibly that the Crumbs group also functions in the generation of cell polarity in the retina.

A separate group, comprised of Par 3 [also known as Bazooka in fly or atypical PKC isotype-specific interacting protein (ASIP)], Par 6, and atypical PKC (zeta and Lambda), colocalizes with the Crumbs group (Izumi et al. 1998; Joberty et al. 2000; Lin et al. 2000; Johansson et al. 2000; Hurd et al. 2003), forming a ternary complex at an early state of epithelial cell development. Presently, there is debate about how the complex is brought together (Suzuki et al. 2001; Nagai-Tamai et al. 2002). Studies of Joberty et al. (Joberty et al. 2000) and Lin et al. (Lin et al. 2000) indicate that Par 3 directly binds Par 6 through a homotypic PDZ interaction that involves the first PDZ of Par 3 and the single PDZ domain of Par 6. In contrast, Nagai-Tami et al. suggested that Par 3 and Par 6 are indirectly brought together through a simultaneous interaction with atypical Protein Kinase C (either lambda or zeta PKC) (Nagai-Tamai et al. 2002). In any case, proper formation and activa-

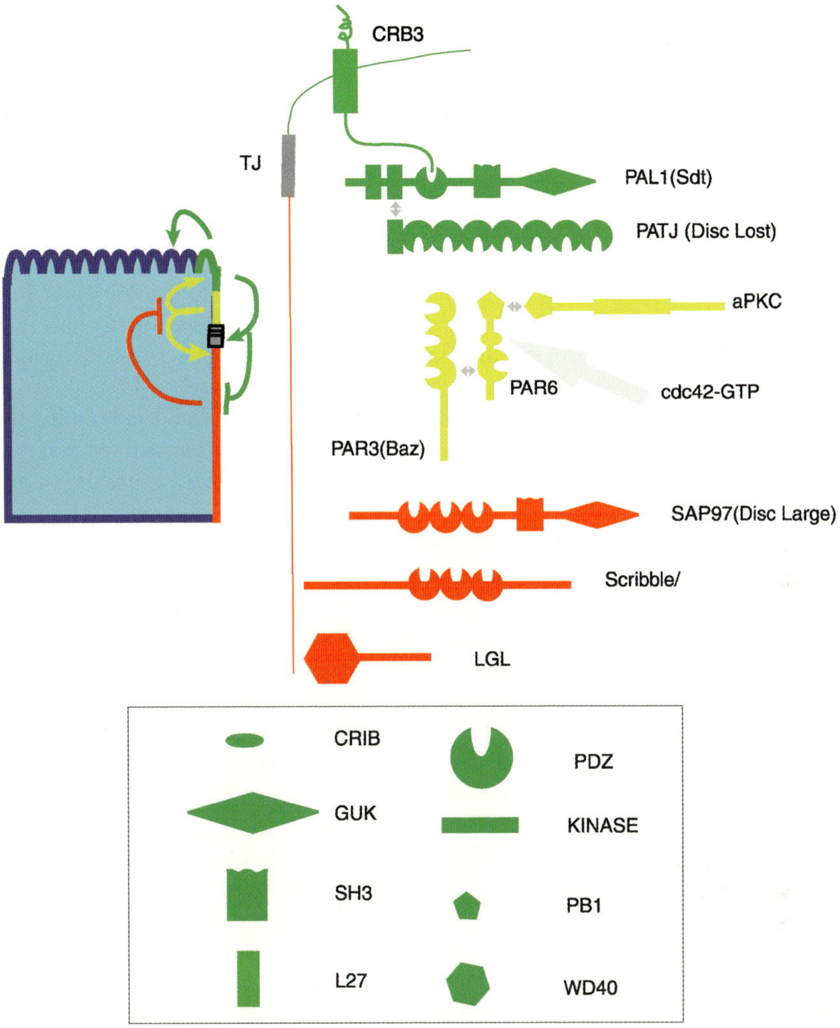

Fig. 4 Three groups of gene products involved in the formation of the epithelial cell–cell junction and the apical-basolateral membrane boundary. Par 3 (*Baz*ooka)/PAR 6/aPKC, form a ternary complex that participates in the initiation of the polarization process, orchestrating tight junction formation. In addition, the Par3/Par6/aPKC complex stimulates the activity the PATJ (Disc Lost)/PAL1 (*Start*dust) /CRB3 (*Crumbs*3) complex, located near the apex of the cell. The apically-oriented complex plays roles in generation of the tight junction and the formation of the apical membrane. A separate group recruited to the lateral membrane is comprised of SAP97 (Disc large)/Scribble/*L*ethal *G*iant *L*avare. This complex antagonizes the function of Par3/Par6/aPKC, repressing apical membrane formation and thereby defining the lateral aspect of the cell. Activities of the complexes are summarized on the *left*. *Green* represents the PATJ (Disc Lost)/ PAL1 (*Start*dust) /CRB3 (*Crumbs*3) complex; *yellow* is the Par3/Bar6/aPKC complex, *red* is SAP97/Scribble/LGL (*arrows*, positive regulation; *bars*, negative regulation). Below is a summary of the functional domains present in members of each complex, CRIB is a Cdc42/Rac interactive binding site; GUK is a guanylate kinase homology domain, SH3 is a Src homology 3 domain; L27 is a *Lin2*/*Lin7* homology domain; PDZ is a protein–protein interaction domain first described in PSD-95, Dlg, and ZO-1/2; the PB1 domain is protein–protein interaction domain named after the first proteins *P*hox and *B*em*1*p that it was found in; WD40 are repeating motifs that form a stable propeller-like protein–protein interaction platform

tion of the complex is necessary for initiation of the polarization process (Joberty et al. 2000; Yamanaka et al. 2001). Indeed, overexpression of Par 3, Par 6 (Joberty et al. 2000), or dominant negative aPKC (Suzuki et al. 2001) disrupts tight junction biogenesis and causes loss of apical-basolateral polarity in MDCK cells.

Importantly, orthologs of these proteins in *C. elegans* were first shown to function in the polarization of mother cells before asymmetric cell division (Kemphues 2000) (hence, Par for Partitioning defective) as well as in the development of epithelial cell polarity in the fly (Kemphues 2000). The implication is that Par 3-Par 6-aPKC is a conserved system for developing and organizing cellular asymmetry in a variety of diverse systems. In the case of mammalian epithelial development, recent studies indicate that the Par 3-Par 6-aPKC complex may physically interact with the more epithelial-specific Crumbs group via a regulated interaction between the mammalian Stardust ortholog, PALS1, and also Par 6 (Hurd et al. 2003).

Understanding just how the Par3-Par6-aPKC complex is regulated has been the focus of extensive investigation. Recent studies indicate that Par 6-aPKC is an important downstream effector of Cdc42 and Rac 1. This would potentially provide a link between the Rho-type small G-proteins in polarization (Takaishi et al. 1997; Jou and Nelson 1998; Kim et al. 2000; Noren et al. 2001), and tight junction formation. The active, GTP-bound forms of Cdc42 specifically bind to the Cdc42/Rac-interactive binding (CRIB) domain of Par 6 (Suzuki et al. 2001; Joberty et al. 2000; Lin et al. 2000). Upon Cdc42-association, Par 6 appears to undergo a conformational change, altering the interaction and activity of two binding partners. First, it markedly enhances the binding affinity for PALS1, presumably to help recruit Par 6 and associated proteins to the Crumbs complex on the apical side of the cell (Hurd et al. 2003). Second, Cdc42 association releases aPKC from Par 6-dependent suppression, and thereby enhances protein kinase activity (Yamanaka et al. 2001). While the identity of all relevant aPKC effectors remains to be determined, it seems likely that junctional assembly is controlled, at least in part, by the Cdc-42-dependent activation of aPKC and phosphorylation of several key building blocks (Suzuki et al. 2001). The recent observation that phosphorylation of Par 3, the first-known substrate of aPKC, destabilizes aPKC binding to PAR3 and frees the kinase to phosphorylate other critical effectors certainly supports this idea (Nagai-Tamai et al. 2002). It remains to be seen if phosphorylation also modifies the interactions of Par 3 with its binding partners such as the immunoglobulin (Ig)-like junctional adhesion molecules JAM-1 (Itoh et al. 2001; Ebnet et al. 2001) and nexin (Takekuni et al. 2003). Such a mechanism would allow a cdc42-dependent recruitment of the Par 3-Par-6-aPKC complex to the forming junction while simultaneously establishing contact with the apical Crumbs group. Hopefully, future studies will provide further insight into the dynamic nature of Cdc42/Rac1-Par 6-Par 3-aPKC interactions, and a more complete understanding of how the complex orchestrates tight junction assembly.

The third group of molecules has largely been defined by genetic studies in *Drosophila* (reviewed in Tepass et al. 2003). This group consists of Discs Large, a MAGUK protein, Scribble, a LAP protein [for *L*eucine Repeats *a*nd *P*DZ (Santoni et al. 2002)], and Lethal Giant Larvae (Bilder et al. 2000; Bilder and Perrimon 2000). It localizes near the plasmalemma at a region that is basal to the Crumbs and Bazooka (Par) complexes. Although it is unknown if these proteins actually form a ternary complex, the localization and behavior of each individual protein is mutually dependent on the expression of the others, as might be expected if they do physically interact. Null mutations in the Discs Large group cause a loss of the epithelial junction in the fly, similar to the Crumbs and Bazooka groups. However, unlike the null mutations in members of the more apical complexes, which inhibit or strongly reduce apical membrane development, null mutations in the Discs Large group enlarge the apical membrane domain (Bilder et al. 2000; Bilder and Perrimon 2000). Consequently, the Discs Large group is thought to restrict apical membrane development and thereby help define the basolateral membrane. While it remains to be determined if the process is recapitulated in mammalian epithelia, the mammalian orthologs of Discs Large (SAP97) (Wu et al. 1998; Reuver and Garner 1998), Scribble (Nakagawa and Huibregtse 2000), and Lethal Giant Larvae (Musch et al. 2002) (see section entitled "SNARE" for alternative or parallel functions of LGL) at least localize along the lateral membrane of MDCK cells.

Exciting genetic studies in *Drosophila* now provide strong evidence that the three disparate groups operate within a single regulatory hierarchy (Tanentzapf and Tepass 2003; Bilder et al. 2003). According to this model, Baz-Par 6-aPKC lies at the top of the chain and initiates junctional assembly and apical polarity. The Discs Large-Scribble-Lethal Giant Larvae group is then independently recruited to the lateral membrane to antagonize the function of Bazooka-Par 6-aPKC and, thereby repress apical membrane formation. Apical recruitment of Crumbs-Stardust-Discs Lost by the Bazooka-Par 6-aPKC then counters the negative function of the Discs Large-Scribble-Lethal Giant Larvae group. By the time a tie has been called in this molecular tug-of-war match between the complexes, a junction is formed and apicol-basal polarity is established. It will be interesting to see if this intriguing mechanism is as well conserved in mammalian epithelia as the polarity complexes.

Summary

Exciting discoveries in the last decade have cast light onto the fundamental mechanisms that underlie polarized trafficking in epithelial cells. It is now clear that epithelial cell membrane asymmetry is achieved by a combination of intracellular sorting operations, vectorial delivery mechanisms and plasmalemma-specific fusion and retention processes. Superposed on this basic conceptual framework, a list of the parts that are required to carry out these functions has begun to be assembled. Indeed, a few well-defined signals that specify polarized segregation, sorting, and retention commands have, now, been described. The intracellular machineries that decode and act on these signals are beginning to be described. In addition, the nature of the molecules that associate with intracellular trafficking vesicles to coordinate polarized delivery, tethering, docking, and fusion are also becoming understood. Combined with the new technologies in live-cell fluorescent microscopy, which permit direct real-time visualization of trafficking processes, pictures of polarization mechanisms are coming into focus.

There is still much more to be learned, however. Many issues remain unresolved. At present, only a small number of conserved sorting signals have been definitively identified. Even less is known about how the majority of sorting signals are recognized and acted upon. Consider, for example, basolateral membrane-directed traffic in the biosynthetic pathway. Although a number of basolateral proteins appear to be sorted soon after biosynthesis by a AP-1B adaptor-dependent mechanism, many others do not. The nature of the basolateral sorting machineries that process these "clathrin-independent" signals remain unknown. There is an even larger gap for apical membrane proteins. With very few exceptions, the molecular bases for apical membrane sorting remain completely unsolved. Thus, it would seem that the most pressing matter involves developing a complete inventory of polarized sorting signals and the machinery that interacts with them. These discoveries should, in turn, pave the way to elucidate the exact sorting events and pathways.

With the application of powerful new tools in cell biology, genomics, and proteomics, we can expect that many of these gaps will be filled in the near future. As they are, it will be crucial to explore the generality of the sorting mechanisms. It will be equally as important to elucidate how epithelial trafficking processes are regulated in concert with physiological demands. In this regard, studies must continue to extend beyond a few select reporter molecules in model epithelial cells to physiologically relevant proteins in their native cellular environments. In doing so, we might come close to fulfilling the field's ultimate ambition. That is, to predict how a given membrane protein, with its own unique repertoire of sorting signals, is expressed on a polarized locale of any given epithelial cell type, with its own unique and potentially regulated assortment of sorting machinery.

Acknowledgements. This review was supported in part by the by National Institutes of Health Grants DK-54231 and DK-63049 to P.A.W.

References

Adams CL, Nelson WJ, Smith SJ (1996) Quantitative analysis of cadherin-catenin-actin reorganization during development of cell–cell adhesion. J Cell Biol 135:1899–1911

Adams CL, Chen YT, Smith SJ, Nelson WJ (1998) Mechanisms of epithelial cell–cell adhesion and cell compaction revealed by high-resolution tracking of E-cadherin-green fluorescent protein. J Cell Biol 142:1105–1119

Adams ME, Mueller HA, Froehner SC (2001) In vivo requirement of the alpha-syntrophin PDZ domain for the sarcolemmal localization of nNOS and aquaporin-4. J Cell Biol 155:113–122

Ahn J, Mundigl O, Muth TR, Rudnick G, Caplan MJ (1996) Polarized expression of GABA transporters in Madin-Darby canine kidney cells and cultured hippocampal neurons. J Biol Chem 271:6917–6924

Al-Awqati, Q (2003) Terminal differentiation of intercalated cells: the role of hensin. Annu Rev Physiol 65:567–583

Alonso MA, Weissman SM (1987) cDNA cloning and sequence of MAL: a hydrophobic protein associated with human T-cell differentiation. Proc Natl Acad Sci 84 1997–2001

Anderson JM (1996) Cell signalling: MAGUK magic. Curr Biol 6:382–334

Ang AL, Folsch H, Koivisto UM, Pypaert M, Mellman I (2003) The Rab 8 GTPase selectively regulates AP-1B-dependent basolateral transport in polarized Madin-Darby canine kidney cells. J Cell Biol 163:339–350

Apodaca G (2001) Endocytic traffic in polarized epithelial cells: role of the actin and microtubule cytoskeleton. Traffic 2:149–159

Apodaca G, Katz LA, Mostov KE (1994) Receptor-mediated transcytosis of IgA in MDCK cells is via apical recycling endosomes. J Cell Biol 125:67–86

Apodaca G, Cardone MH, Whiteheart SW, Dasgupta BR, Mostov KE (1996) Reconstitution of transcytosis in SLO-permeabilized MDCK cells: existence of an NSF-dependent fusion mechanism with the apical surface of MDCK cells. EMBO J 15:1471–1481

Aroeti B, Kosen PA, Kuntz ID, Cohen FE, Mostov KE (1993) Mutational and secondary structural analysis of the basolateral sorting signal of the polymeric immunoglobulin receptor. J Cell Biol 123:1149–1160

Asakura T, Nakanishi H, Sakisaka T, Takahashi K, Mandai K, Nishimura M, Sasaki T, Takai Y (1999) Similar and differential behaviour between the nectin-afadin-ponsin and cadherin-catenin systems during the formation and disruption of the polarized junctional alignment in epithelial cells. Genes Cells 4:573–581

Bacallao R, Antony C, Dotti C, Karsenti E, Stelzer EH, Simons K (1989) The subcellular organization of Madin-Darby canine kidney cells during the formation of a polarized epithelium. J Cell Biol 109:2817–2832

Bachmann A, Schneider M, Theilenberg E, Grawe F, Knust E (2001) Drosophila Stardust is a partner of Crumbs in the control of epithelial cell polarity. Nature 414:638–643

Balcarova-Stander J, Pfeiffer SE, Fuller SD, Simons K (1984) Development of cell surface polarity in the epithelial Madin-Darby canine kidney (MDCK) cell line. EMBO J 3:2687–2694

Banerjee A, Shih T, Alexander EA, Schwartz JH (1999) SNARE proteins regulate H(+)-ATPase redistribution to the apical membrane in rat renal inner medullary collecting duct cells. J Biol Chem 274:26518–26522

Banerjee A, Li G, Alexander EA, Schwartz JH (2001) Role of SNAP-23 in trafficking of H+-ATPase in cultured inner medullary collecting duct cells. Am J Physiol Cell Physiol 280:C775–C781

Barnard RJ, Morgan A, Burgoyne RD (1997) Stimulation of NSF ATPase activity by alpha-SNAP is required for SNARE complex disassembly and exocytosis. J Cell Biol 139:875–883

Barth AI, Nathke IS, Nelson WJ (1997) Cadherins, catenins and APC protein: interplay between cytoskeletal complexes and signaling pathways. Curr Opin Cell Biol 9:683–690

Bartles JR, Feracci HM, Stieger B, Hubbard AL (1987) Biogenesis of the rat hepatocyte plasma membrane in vivo: comparison of the pathways taken by apical and basolateral proteins using subcellular fractionation. J Cell Biol 105:1241–1251

Benharouga M, Sharma M, So J, Haardt M, Drzymala L, Popov M, Schwapach B, Grinstein S, Du K, Lukacs GL (2003) The role of the C terminus and na+/h+ exchanger regulatory factor in the functional expression of cystic fibrosis transmembrane conductance regulator in nonpolarized cells and epithelia. J Biol Chem 278:22079–22089

Bennett MK, Wandinger-Ness A, Simons K (1988) Release of putative exocytic transport vesicles from perforated MDCK cells. EMBO J 7:4075–4085

Benting JH, Rietveld AG, Simons K (1999) N-Glycans mediate the apical sorting of a GPI-anchored, raft-associated protein in Madin-Darby canine kidney cells. J Cell Biol 146:313–320

Bhat MA, Izaddoost S, Lu Y, Cho KO, Choi KW, Bellen HJ (1999) Discs Lost a novel multi-PDZ domain protein, establishes and maintains epithelial polarity. Cell 96:833–845

Biemesderfer D, Mentone SA, Mooseker M, Hasson T (2002) Expression of myosin VI within the early endocytic pathway in adult and developing proximal tubules. Am J Physiol Renal Physiol 282:F785–F794

Bilder D, Perrimon N (2000) Localization of apical epithelial determinants by the basolateral PDZ protein Scribble. Nature 403:676–680

Bilder D, Li M, Perrimon N (2000) Cooperative regulation of cell polarity and growth by Drosophila tumor suppressors. Science 289:113–116

Bilder D, Schober M, Perrimon N (2003) Integrated activity of PDZ protein complexes regulates epithelial polarity. Nat Cell Biol 5:53–58

Boehm M, Bonifacino JS (2001) Adaptins: the final recount. Mol Biol Cell 12:2907–2920

Boll W, Partin JS, Katz AI, Caplan MJ, Jamieson JD (1991) Distinct pathways for basolateral targeting of membrane and secretory proteins in polarized epithelial cells. Proc Natl Acad Sci 88:8592–8596

Bomsel M, Prydz K, Parton RG, Gruenberg J, Simons K (1989) Endocytosis in filter-grown Madin-Darby canine kidney cells. J Cell Biol 109:3243–3258

Bomsel M, Parton R, Kuznetsov SA, Schroer TA, Gruenberg J (1990) Microtubule- and motor-dependent fusion in vitro between apical and basolateral endocytic vesicles from MDCK cells. Cell 62:719–731

Bonifacino JS, Dell'angelica EC (1999) Molecular bases for the recognition of tyrosine-based sorting signals. J Cell Biol 145:923–926

Bonifacino JS, Traub LM (2003) Signals for sorting of transmembrane proteins to endosomes and lysosomes. Annu Rev Biochem 72:395–447

Borg JP, Straight SW, Kaech SM, De Taddeo-Borg M, Kroon DE, Karnak D, Turner RS, Kim SK, Margolis B (1998) Identification of an evolutionarily conserved heterotrimeric protein complex involved in protein targeting. J Biol Chem 273:31633–31636

Borg JP, Marchetto S, Le Bivic A, Ollendorff V, Jaulin-Bastard F, Saito H, Fournier E, Adelaide J, Margolis B, Birnbaum D (2000) ERBIN: a basolateral PDZ protein that interacts with the mammalian ERBB2/HER2 receptor. Nat Cell Biol 2:407–414

Breitfeld PP, Mckinnon WC, Mostov KE (1990) Effect of nocodazole on vesicular traffic to the apical and basolateral surfaces of polarized MDCK cells. J Cell Biol 111:2365–2373

Breton S, Lisanti MP, Tyszkowski R, Mclaughlin M, Brown D (1998) Basolateral distribution of caveolin-1 in the kidney. Absence from H+-atpase-coated endocytic vesicles in intercalated cells. J Histochem Cytochem 46:205–214

Breton S, Nsumu NN, Galli T, Sabolic I, Smith PJ, Brown D (2000) Tetanus toxin-mediated cleavage of cellubrevin inhibits proton secretion in the male reproductive tract. Am J Physiol Renal Physiol 278:F717–F725

Breton S, Inoue T, Knepper MA, Brown D (2002) Antigen retrieval reveals widespread basolateral expression of syntaxin 3 in renal epithelia. Am J Physiol Renal Physiol 282:F523–F559

Brown A, Muth T, Caplan MJ (2004) The C-terminal tail of the GAT-2 GABA transporter contains a novel motif that plays a role in basolateral targeting. Am J Physiol Cell Physiol 286(5):C1071–1077

Brown DA, London E (1998) Functions of lipid rafts in biological membranes. Annu Rev Cell Dev Biol 14:111–136

Brown DA, Rose JK (1992) Sorting of GPI-anchored proteins to glycolipid-enriched membrane subdomains during transport to the apical cell surface. Cell 68:533–544

Brown DA, Crise B, Rose JK (1989) Mechanism of membrane anchoring affects polarized expression of two proteins in MDCK cells. Science 245:1499–1501

Brown PS, Wang E, Aroeti B, Chapin SJ, Mostov KE, Dunn KW (2000) Definition of distinct compartments in polarized Madin-Darby canine kidney (MDCK) cells for membrane-volume sorting, polarized sorting and apical recycling. Traffic 1:124–140

Bucci C, Wandinger-Ness A, Lutcke A, Chiariello M, Bruni CB, Zerial M (1994) Rab 5a is a common component of the apical and basolateral endocytic machinery in polarized epithelial cells. Proc Natl Acad Sci 91:5061–5065

Bunn RC, Jensen MA, Reed BC (1999) Protein interactions with the glucose transporter binding protein GLUT1CBP that provide a link between GLUT1 and the cytoskeleton. Mol Biol Cell 10:819–832

Butz S, Okamoto M, Sudhof TC (1998) A tripartite protein complex with the potential to couple synaptic vesicle exocytosis to cell adhesion in brain. Cell 94:773–782

Calhoun BC, Lapierre LA, Chew CS, Goldenring JR (1998) Rab 11a redistributes to apical secretory canaliculus during stimulation of gastric parietal cells. Am J Physiol 275:C163–C170

Cao TT, Deacon HW, Reczek D, Bretscher A, Von Zastrow M (1999) A kinase-regulated PDZ-domain interaction controls endocytic sorting of the beta2-adrenergic receptor. Nature 401:286–290

Caplan MJ (1997) Membrane polarity in epithelial cells: protein sorting and establishment of polarized domains. Am J Physiol 272:F425–429

Caplan MJ, Anderson HC, Palade GE, Jamieson JD (1986) Intracellular sorting and polarized cell surface delivery of (Na+,K+)ATPase, an endogenous component of MDCK cell basolateral plasma membranes. Cell 46(4):623–631

Carroll KS, Hanna J, Simon I, Krise J, Barbero P, Pfeffer SR (2001) Role of Rab 9 GTPase in facilitating receptor recruitment by TIP47. Science 292:1373–1376

Casanova JE, Apodaca G, Mostov KE (1991) An autonomous signal for basolateral sorting in the cytoplasmic domain of the polymeric immunoglobulin receptor. Cell 66:65–75

Casanova JE, Wang X, Kumar R, Bhartur SG, Navarre J, Woodrum JE, Altschuler Y, Ray GS, Goldenring JR (1999) Association of Rab 25 and Rab 11a with the apical recycling system of polarized Madin-Darby canine kidney cells. Mol Biol Cell 10:47–61

Cereijido M, Valdes J, Shoshani L, Contreras RG (1998) Role of tight junctions in establishing and maintaining cell polarity. Annu Rev Physiol 60:161–177

Cereijido M, Shoshani L, Contreras RG (2000) Molecular physiology and pathophysiology of tight junctions. I. Biogenesis of tight junctions and epithelial polarity. Am J Physiol Gastrointest Liver Physiol 279:G477–G482

Charron AJ, Bacallao RL, Wandinger-Ness A (2000) ADPKD: a human disease altering Golgi function and basolateral exocytosis in renal epithelia. Traffic 1:675-686

Chen YA, Scheller RH (2001) SNARE-mediated membrane fusion. Nat Rev Mol Cell Biol 2:98–106

Cheng J, Moyer BD, Milewski M, Loffing J, Ikeda M, Mickle JE, Cutting GR, Li M, Stanton BA, Guggino WB (2002) A Golgi-associated PDZ domain protein modulates cystic fibrosis transmembrane regulator plasma membrane expression. J Biol Chem 277:3520–3529

Cheong KH, Zacchetti D, Schneeberger EE, Simons K (1999) VIP17/MAL a lipid raft-associated protein, is involved in apical transport in MDCK cells. Proc Natl Acad Sci 96:6241–6248

Cho EA, Patterson LT, Brookhiser WT, Mah S, Kintner C, Dressler GR (1998) Differential expression and function of cadherin-6 during renal epithelium development. Development 125:803–812

Christoforidis S, Mcbride HM, Burgoyne RD, Zerial M (1999) The Rab 5 effector EEA1 is a core component of endosome docking. Nature 397:621–625

Chuang JZ, Sung CH (1998) The cytoplasmic tail of rhodopsin acts as a novel apical sorting signal in polarized MDCK cells. J Cell Biol 142:1245–1556

Cohen AR, Woods DF, Marfatia SM, Walther Z, Chishti AH, Anderson JM, Wood DFW (1998) Human CASK/LIN-2 binds syndecan-2 and protein 4.1 and localizes to the basolateral membrane of epithelial cells [published erratum appears in J Cell Biol 1998 Aug 24;142(4):following 1156]. J Cell Biol 142:129–138

Daniels DL, Cohen AR, Anderson JM, Brunger AT (1998) Crystal structure of the hCASK PDZ domain reveals the structural basis of class II PDZ domain target recognition. Nat Struct Biol 5:317–325

Delgrossi MH, Breuza L, Mirre C, Chavrier P, Le Bivic A (1997) Human syntaxin 3 is localized apically in human intestinal cells. J Cell Sci 110:2207–2214

Den Hollander AI, Ten Brink JB, De Kok YJ, Van Soest S, Van Den Born LI, Van Driel MA, Van De Pol DJ, Payne AM, Bhattacharya SS, Kellner U, Hoyng CB, Westerveld A, Brunner HG, Bleeker-Wagemakers EM, Deutman AF, Heckenlively JR, Cremers FP, Bergen AA (1999) Mutations in a human homologue of Drosophila crumbs cause retinitis pigmentosa (RP12) Nat Genet 23:217–221

Duman JG, Tyagarajan K, Kolsi MS, Moore HP, Forte JG (1999) Expression of rab11a N124I in gastric parietal cells inhibits stimulatory recruitment of the H+-K+-ATPase. Am J Physiol 277:C361–C372

Dunbar LA, Aronson P, Caplan MJ (2000) A transmembrane segment determines the steady-state localization of an ion-transporting adenosine triphosphatase. J Cell Biol 148:769–778

Dunn KW, Mcgraw TE, Maxfield FR (1989) Iterative fractionation of recycling receptors from lysosomally destined ligands in an early sorting endosome. J Cell Biol 109:3303–3314

Ebnet K, Suzuki A, Horikoshi Y, Hirose T, Meyer Zu Brickwedde MK, Ohno S, Vestweber D (2001) The cell polarity protein ASIP/PAR-3 directly associates with junctional adhesion molecule (JAM) EMBO J 20:3738–3748

Ekblom P, Vestweber D, Kemler R (1986) Cell–matrix interactions and cell adhesion during development. Annu Rev Cell Dev Biol 2:27–47

Evans WH (1980) A biochemical dissection of the functional polarity of the plasma membrane of the hepatocyte. Biochim Biophys Acta 604:27–64

Fasshauer D, Sutton RB, Brunger AT, Jahn R (1998) Conserved structural features of the synaptic fusion complex: SNARE proteins reclassified as Q- and R-SNAREs. Proc Natl Acad Sci 95:15781–15786

Fernandez-Larrea J, Merlos-Suarez A, Urena JM, Baselga J, Arribas J (1999) A role for a PDZ protein in the early secretory pathway for the targeting of proTGF-alpha to the cell surface. Mol Cell 3:423–433

Fialka I, Steinlein P, Ahorn H, Bock G, Burbelo PD, Haberfellner M, Lottspeich F, Paiha K, Pasquali C, Huber LA (1999) Identification of syntenin as a protein of the apical early endocytic compartment in Madin-Darby canine kidney cells. J Biol Chem 274:26233–26239

Fiedler K, Lafont F, Parton RG, Simons K (1995) Annexin XIIIb: a novel epithelial specific annexin is implicated in vesicular traffic to the apical plasma membrane. J Cell Biol 128:1043–1053

Finger FP, Hughes TE, Novick P (1998) Sec 3p is a spatial landmark for polarized secretion in budding yeast. Cell 92:559–571

Foletti DL, Prekeris R, Scheller RH (1999) Generation and maintenance of neuronal polarity: mechanisms of transport and targeting. Neuron 23:641–644

Folsch H, Ohno H, Bonifacino JS, Mellman I (1999) A novel clathrin adaptor complex mediates basolateral targeting in polarized epithelial cells. Cell 99:189–198

Folsch H, Pypaert M, Schu P, Mellman I (2001) Distribution and function of AP-1 clathrin adaptor complexes in polarized epithelial cells. J Cell Biol 152:595–606

Folsch H, Pypaert M, Maday S, Pelletier L, Mellman I (2003) The AP-1A and AP-1B clathrin adaptor complexes define biochemically and functionally distinct membrane domains. J Cell Biol 163:351–362

Frank AE, Wingo CS, Andrews PM, Ageloff S, Knepper MA, Weiner ID (2002) Mechanisms through which ammonia regulates cortical collecting duct net proton secretion. Am J Physiol Renal Physiol 282:F1120–F1128

Franki N, Macaluso F, Gao Y, Hays RM (1995) Vesicle fusion proteins in rat inner medullary collecting duct and amphibian bladder. Am J Physiol 268:C792–C797

Fukuhara A, Irie K, Nakanishi H, Takekuni K, Kawakatsu T, Ikeda W, Yamada A, Katata T, Honda T, Sato T, Shimizu K, Ozaki H, Horiuchi H, Kita T, Takai Y (2002) Involvement of nectin in the localization of junctional adhesion molecule at tight junctions. Oncogene 21:7642–7655

Futter CE, Gibson A, Allchin EH, Maxwell S, Ruddock LJ, Odorizzi G, Domingo D, Trowbridge IS, Hopkins CR (1998) In polarized MDCK cells basolateral vesicles arise from clathrin-gamma-adaptin-coated domains on endosomal tubules. J Cell Biol 141:611–623

Galli T, Zahraoui A, Vaidyanathan VV, Raposo G, Tian JM, Karin M, Niemann H, Louvard D (1998) A novel tetanus neurotoxin-insensitive vesicle-associated membrane protein in SNARE complexes of the apical plasma membrane of epithelial cells. Mol Biol Cell 9:1437–1438

Gan Y, Mcgraw TE, Rodriguez-Boulan E (2002) The epithelial-specific adaptor AP1B mediates post-endocytic recycling to the basolateral membrane. Nat Cell Biol 4:605–609

Geffen I, Fuhrer C, Leitinger B, Weiss M, Huggel K, Griffiths G, Spiess M (1993) Related signals for endocytosis and basolateral sorting of the asialoglycoprotein receptor. J Biol Chem 268:20772–20777

Gisler SM, Stagljar I, Traebert M, Bacic D, Biber J, Murer H (2001) Interaction of the type IIa Na/Pi cotransporter with PDZ proteins. J Biol Chem 276:9206–9213

Goldenring JR, Shen KR, Vaughan HD, Modlin IM (1993) Identification of a small GTP-binding protein, Rab 25, expressed in the gastrointestinal mucosa, kidney, and lung. J Biol Chem 268:18419–18422

Goldenring JR, Smith J, Vaughan HD, Cameron P, Hawkins W, Navarre J (1996) Rab 11 is an apically located small GTP-binding protein in epithelial tissues. Am J Physiol 270:G515–G525

Gomperts SN (1996) Clustering membrane proteins: It's all coming together with the PSD-95/SAP90 protein family. [Review] [20 refs]. Cell 84:659–662

Gottardi CJ, Dunbar LA, Caplan MJ (1995) Biotinylation and assessment of membrane polarity: caveats and methodological concerns. Am J Physiol 268:F285–F295

Gottlieb TA, Ivanov IE, Adesnik M, Sabatini DD (1993) Actin microfilaments play a critical role in endocytosis at the apical but not the basolateral surface of polarized epithelial cells. J Cell Biol 120:695–710

Gouraud S, Laera A, Calamita G, Rossetto O, Montecucco C, Rosenthal W, Svelto M, Valenti G (2001) Functional involvement of the SNARE machinery in cAMP induced AQP-2 targeting to the apical membrane in renal Epithelilia cells. J Am Soc Nephrol12:A0302 (abstract)

Grantham JJ, Uchic M, Cragoe EJ, Jr, Kornhaus J, Grantham JA, Donoso V, Mangoo-Karim R, Evan A, Mcateer J (1989) Chemical modification of cell proliferation and fluid secretion in renal cysts. Kidney Int 35:1379–1389

Grawe F, Wodarz A, Lee B, Knust E, Skaer H (1996) The Drosophila genes crumbs and stardust are involved in the biogenesis of adherens junctions. Development 122:951–999

Griffiths G, Pfeiffer S, Simons K, Matlin K (1985) Exit of newly synthesized membrane proteins from the trans cisterna of the Golgi complex to the plasma membrane. J Cell Biol 101:949–964

Grindstaff KK, Bacallao RL, Nelson WJ (1998a) Apiconuclear organization of microtubules does not specify protein delivery from the trans-Golgi network to different membrane domains in polarized epithelial cells. Mol Biol Cell 9:685–699

Grindstaff KK, Yeaman C, Anandasabapathy N, Hsu SC, Rodriguez-Boulan E, Scheller RH, Nelson WJ (1998b) Sec 6/8 complex is recruited to cell–cell contacts and specifies transport vesicle delivery to the basal-lateral membrane in epithelial cells. Cell 93:731–740

Gu HH, Wu X, Giros B, Caron MG, Caplan MJ, Rudnick G (2001) The NH(2)-terminus of Norepinephrine transporter contains a basolateral localization signal for epithelial cells. Mol Biol Cell 12:3797–3807

Guo W, Roth D, Walch-Solimena C, Novick P (1999) The exocyst is an effector for Sec 4p, targeting secretory vesicles to sites of exocytosis. EMBO J 18:1071–1080

Guo W, Sacher M, Barrowman J, Ferro-Novick S, Novick P (2000) Protein complexes in transport vesicle targeting. Trends Cell Biol 10:251–255

Hales CM, Vaerman JP, Goldenring JR (2002) Rab 11 family interacting protein 2 associates with Myosin Vb and regulates plasma membrane recycling. J Biol Chem 277:50415–50421

Hamm-Alvarez SF, Sheetz MP (1998) Microtubule-dependent vesicle transport: modulation of channel and transporter activity in liver and kidney. Physiol Rev 78:1109–1209

Hammer JA, Wu XS (2002) Rabs grab motors: defining the connections between Rab GTPases and motor proteins. Curr Opin Cell Biol 14:69–75

Hannan LA, Edidin M (1996) Traffic, polarity, and detergent solubility of a glycosylphosphatidylinositol-anchored protein after LDL-deprivation of MDCK cells. J Cell Biol 133:1265–1276

Hara-Kuge S, Ohkura T, Ideo H, Shimada O, Atsumi S, Yamashita K (2002) Involvement of VIP36 in intracellular transport and secretion of glycoproteins in polarized Madin-Darby canine kidney (MDCK) cells. J Biol Chem 277:16332–16339

Harris BZ, Hillier BJ, Lim WA (2001) Energetic determinants of internal motif recognition by PDZ domains. Biochim Biophys Acta 40:5921–5930

Hata Y, Slaughter CA, Sudhof TC (1993) Synaptic vesicle fusion complex contains unc-18 homologue bound to syntaxin. Nature 366:347–351

Hernando N, Karim-Jimenez Z, Biber J, Murer H (2001) Molecular determinants for apical expression and regulatory membrane retrieval of the type IIa Na/Pi cotransporter. Kidney Int 60:431–435

Hernando N, Deliot N, Gisler SM, Lederer E, Weinman EJ, Biber J, Murer H (2002) PDZ-domain interactions and apical expression of type IIa Na/P(i) cotransporters. Proc Natl Acad Sci 99:11957–11962

Herzlinger DA, Ojakian GK (1984) Studies on the development and maintenance of epithelial cell surface polarity with monoclonal antibodies. J Cell Biol 98:1777–1787

Hillier BJ, Christopherson KS, Prehoda KE, Bredt DS, Lim WA (1999) Unexpected modes of PDZ domain scaffolding revealed by structure of nNOS-syntrophin complex. Science 284:812–815

Honda T, Shimizu K, Kawakatsu T, Yasumi M, Shingai T, Fukuhara A, Ozaki-Kuroda K, Irie K, Nakanishi H, Takai Y (2003) Antagonistic and agonistic effects of an extracellular fragment of nectin on formation of E-cadherin-based cell–cell adhesion. Genes Cells 8:51–63

Hsu SC, Ting AE, Hazuka CD, Davanger S, Kenny JW, Kee Y, Scheller RH (1996) The mammalian brain rsec6/8 complex. Neuron 17:1209–1219

Hsu SC, Hazuka CD, Foletti DL, Scheller RH (1999) Targeting vesicles to specific sites on the plasma membrane: the role of the sec6/8 complex. Trends Cell Biol 9:150-153

Huber LA, Pimplikar S, Parton RG, Virta H, Zerial M, Simons K (1993) Rab 8 a small GTPase involved in vesicular traffic between the TGN and the basolateral plasma membrane. J Cell Biol 123:35–45

Huber O, Bierkamp C, Kemler R (1996) Cadherins and catenins in development. Curr Opin Cell Biol 8:685–691

Hughson EJ, Hopkins CR (1990) Endocytic pathways in polarized Caco-2 cells: identification of an endosomal compartment accessible from both apical and basolateral surfaces. J Cell Biol 110:337–348

Hull RN, Cherry WR, Weaver GW (1976) The origin and characteristics of a pig kidney cell strain, LLC-PK. In Vitro 12:670–677

Hunt CA, Schenker LJ, Kennedy MB (1996) PSD-95 is associated with the postsynaptic density and not with the presynaptic membrane at forebrain synapses. J Neurosci 16:1380–1388

Hunziker W, Fumey C (1994) A di-leucine motif mediates endocytosis and basolateral sorting of macrophage IgG Fc receptors in MDCK cells. EMBO J 13:2963–2967

Hunziker W, Peters PJ (1998) Rab 17 localizes to recycling endosomes and regulates receptor-mediated transcytosis in epithelial cells. J Biol Chem 273:15734–15741

Hurd TW, Gao L, Roh MH, Macara IG, Margolis B (2003) Direct interaction of two polarity complexes implicated in epithelial tight junction assembly. Nat Cell Biol 5:137–142

Ibaraki K, Horikawa HP, Morita T, Mori H, Sakimura K, Mishina M, Saisu H, Abe T (1995) Identification of four different forms of syntaxin 3. Biochem Biophys Res Commun 211:997–1005

Ikonen E (2001) Roles of lipid rafts in membrane transport. Curr Opin Cell Biol 13:470–477

Ikonen E, Tagaya M, Ullrich O, Montecucco C, Simons K (1995) Different requirements for NSF, SNAP, and Rab proteins in apical and basolateral transport in MDCK cells. Cell 81:571–580

Inoue T, Nielsen S, Mandon B, Terris J, Kishore BK, Knepper MA (1998) SNAP-23 in rat kidney: colocalization with aquaporin-2 in collecting duct vesicles. Am J Physiol 275:F752–F760

Itoh M, Sasaki H, Furuse M, Ozaki H, Kita T, Tsukita S (2001) Junctional adhesion molecule (JAM) binds to PAR-3: a possible mechanism for the recruitment of PAR-3 to tight junctions. J Cell Biol 154:491–497

Izumi Y, Hirose T, Tamai Y, Hirai S, Nagashima Y, Fujimoto T, Tabuse Y, Kemphues KJ, Ohno S (1998) An atypical PKC directly associates and colocalizes at the epithelial tight junction with ASIP a mammalian homologue of Caenorhabditis elegans polarity protein PAR-3. J Cell Biol 143:95–106

Jacob R, Alfalah M, Grunberg J, Obendorf M, Naim HY (2000) Structural determinants required for apical sorting of an intestinal brush-border membrane protein. J Biol Chem 275:6566–6572

Jacob R, Heine M, Alfalah M, Naim HY (2003) Distinct cytoskeletal tracks direct individual vesicle populations to the apical membrane of epithelial cells. Curr Biol 13:607–612

Jacob R, Heine M, Eikemeyer J, Frerker N, Zimmer KP, Rescher U, Gerke V, Naim HY (2004) Annexin II is required for apical transport in polarized epithelial cells. J Biol Chem 279 (5):3680–3684

Jahn R, Sudhof TC (1999) Membrane fusion and exocytosis. Annu Rev Biochem 68:863–911

Jesaitis LA, Goodenough DA (1994) Molecular characterization and tissue distribution of ZO-2 a tight junction protein homologous to ZO-1 and the Drosophila discs-large tumor suppressor protein. J Cell Biol 124:949–961

Jo I, Harris HW, Amendt-Raduege AM, Majewski RR, Hammond TG (1995) Rat kidney papilla contains abundant synaptobrevin protein that participates in the fusion of antidiuretic hormone-regulated water channel-containing endosomes in vitro. Proc Natl Acad Sci 92:1876–1880

Joberty G, Petersen C, Gao L, Macara IG (2000) The cell-polarity protein Par6 links Par3 and atypical protein kinase C to Cdc42. Nat Cell Biol 2:531–559

Johansson A, Driessens M, Aspenstrom P (2000) The mammalian homologue of the Caenorhabditis elegans polarity protein PAR-6 is a binding partner for the Rho GTPases Cdc42 and Rac1. J Cell Sci 113:3267–3275

Jou TS, Nelson WJ (1998) Effects of regulated expression of mutant RhoA and Rac1 small GTPases on the development of epithelial (MDCK) cell polarity. J Cell Biol 142:85–100

Jou TS, Stewart DB, Stappert J, Nelson WJ, Marrs JA (1995) Genetic and biochemical dissection of protein linkages in the cadherin-catenin complex. Proc Natl Acad Sci 92:5067–5071

Kachinsky AM, Froehner SC, Milgram SL (1999) A PDZ-containing scaffold related to the dystrophin complex at the basolateral membrane of epithelial cells. J Cell Biol 145:391–402

Kaech SM, Whitfield CW, Kim SK (1998) The LIN-2/LIN-7/LIN-10 complex mediates basolateral membrane localization of the C. elegans EGF receptor LET-23 in vulval epithelial cells. Cell 94:761–771

Kamberov E, Makarova O, Roh M, Liu A, Karnak D, Straight S, Margolis B (2000) Molecular cloning and characterization of Pals, proteins associated with mLin-7. J Biol Chem 275:11425–11431

Karim-Jimenez Z, Hernando N, Biber J, Murer H (2001) Molecular determinants for apical expression of the renal type IIa Na+/Pi-cotransporter. Pflugers Arch Eur J Physiol 442:782–790

Keller P, Simons K (1998) Cholesterol is required for surface transport of influenza virus hemagglutinin. J Cell Biol 140:1357–1367

Keller P, Toomre D, Diaz E, White J, Simons K (2001) Multicolour imaging of post-Golgi sorting and trafficking in live cells. Nat Cell Biol 3:140–149

Kemphues K (2000) PARsing embryonic polarity. Cell 101:345–348

Kim E, Niethammer M, Rothschild A, Jan YN, Sheng M (1995) Clustering of Shaker-type K+ channels by interaction with a family of membrane-associated guanylate kinases. Nature 378:85–88

Kim SH, Li Z, Sacks DB (2000) E-cadherin-mediated cell–cell attachment activates Cdc42. J Biol Chem 275:36999–37005

Klein G, Langegger M, Timpl R, Ekblom P (1988) Role of laminin A chain in the development of epithelial cell polarity. Cell 55:331–341

Knight A, Hughson E, Hopkins CR, Cutler DF (1995) Membrane protein trafficking through the common apical endosome compartment of polarized Caco-2 cells. Mol Biol Cell 6:597–610

Knust E, Bossinger O (2002) Composition and formation of intercellular junctions in epithelial cells. Science 298:1955-1959

Knust E, Tepass U, Wodarz A (1993) crumbs and stardust, two genes of Drosophila required for the development of epithelial cell polarity. Development (Suppl):261–280

Koob R, Zimmermann M, Schoner W, Drenckhahn D (1988) Colocalization and coprecipitation of ankyrin and Na+,K+-ATPase in kidney epithelial cells. Eur J Cell Biol 45:230–237

Kreis TE (1992) Regulation of vesicular and tubular membrane traffic of the Golgi complex by coat proteins. Curr Opin Cell Biol 4:609–615

Kreitzer G, Marmorstein A, Okamoto P, Vallee R, Rodriguez-Boulan E (2000) Kinesin and dynamin are required for post-Golgi transport of a plasma-membrane protein. Nat Cell Biol 2:125–117

Kreitzer G, Schmoranzer J, Low SH, Li X, Gan Y, Weimbs T, Simon SM, Rodriguez-Boulan E (2003) Three-dimensional analysis of post-Golgi carrier exocytosis in epithelial cells. Nat Cell Biol 5:126–136

Kroschewski R, Hall A, Mellman I (1999) Cdc42 controls secretory and endocytic transport to the basolateral plasma membrane of MDCK cells. Nat Cell Biol 1:8–13

Kundu A, Avalos RT, Sanderson CM, Nayak DP (1996) Transmembrane domain of influenza virus neuraminidase a type II protein, possesses an apical sorting signal in polarized MDCK cells. J Virol 70:6508–6515

Ladinsky MS, Kremer JR, Furcinitti PS, Mcintosh JR, Howell KE (1994) HVEM tomography of the trans-Golgi network: structural insights and identification of a lace-like vesicle coat. J Cell Biol 127:29–38

Ladinsky MS, Mastronarde DN, Mcintosh JR, Howell KE, Staehelin LA (1999) Golgi structure in three dimensions: functional insights from the normal rat kidney cell. J Cell Biol 144:1135–1149

Ladinsky MS, Wu CC, Mcintosh S, Mcintosh JR, Howell KE (2002) Structure of the Golgi and distribution of reporter molecules at 20 degrees C reveals the complexity of the exit compartments. Mol Biol Cell 13:2810–2825

Lafont F, Burkhardt JK, Simons K (1994) Involvement of microtubule motors in basolateral and apical transport in kidney cells. Nature 372:801–803

Lafont F, Lecat S, Verkade P, Simons K (1998) Annexin XIIIb associates with lipid microdomains to function in apical delivery. J Cell Biol 142:1413–1427

Lafont F, Verkade P, Galli T, Wimmer C, Louvard D, Simons K (1999) Raft association of SNAP receptors acting in apical trafficking in Madin-Darby canine kidney cells. Proc Natl Acad Sci 96:3734–3738

Lahtinen U, Honsho M, Parton RG, Simons K, Verkade P (2003) Involvement of caveolin-2 in caveolar biogenesis in MDCK cells. FEBS Lett 538:85–88

Langford GM (2002) Myosin-V a versatile motor for short-range vesicle transport. Traffic 3:859–865

Lapierre LA, Kumar R, Hales CM, Navarre J, Bhartur SG, Burnette JO, Provance DW, Jr, Mercer JA, Bahler M, Goldenring JR (2001) Myosin vb is associated with plasma membrane recycling systems. Mol Biol Cell 12:1843–1857

Le Gall AH, Powell SK, Yeaman CA, Rodriguez-Boulan E (1997) The neural cell adhesion molecule expresses a tyrosine-independent basolateral sorting signal. J Biol Chem 272:4559–4567

Le Maout S, Welling PA, Brejon M, Olsen O, Merot J (2001) Basolateral membrane expression of a K+ channel, Kir 2.3, is directed by a cytoplasmic COOH-terminal domain. Proc Natl Acad Sci 98:10475–10480

Lee S, Fan S, Makarova O, Straight S, Margolis B (2002) A novel and conserved protein-protein interaction domain of mammalian Lin-2/CASK binds and recruits SAP97 to the lateral surface of epithelia. Mol Cell Biol 22:1778–1791

Leighton J, Brada Z, Estes LW, Justh G (1969) Secretory activity and oncogenicity of a cell line (MDCK) derived from canine kidney. Science 163:472–473

Leung SM, Ruiz WG, Apodaca G (2000) Sorting of membrane and fluid at the apical pole of polarized Madin-Darby canine kidney cells. Mol Biol Cell 11:2131–2150

Li X, Low SH, Miura M, Weimbs T (2002) SNARE expression and localization in renal epithelial cells suggest mechanism for variability of trafficking phenotypes. Am J Physiol Renal Physiol 283:F1111–F1122

Lin D, Edwards AS, Fawcett JP, Mbamalu G, Scott JD, Pawson T (2000) A mammalian PAR-3-PAR-6 complex implicated in Cdc42/Rac1 and aPKC signalling and cell polarity. Nat Cell Biol 2:540–547

Lin S, Naim HY, Rodriguez AC, Roth MG (1998) Mutations in the middle of the transmembrane domain reverse the polarity of transport of the influenza virus hemagglutinin in MDCK epithelial cells. J Cell Biol 142:51–57

Lipardi C, Mora R, Colomer V, Paladino S, Nitsch L, Rodriguez-Boulan E, Zurzolo C (1998) Caveolin transfection results in caveolae formation but not apical sorting of glycosylphosphatidylinositol (GPI)-anchored proteins in epithelial cells. J Cell Biol 140:617–626

Lipardi C, Nitsch L, Zurzolo C (2000) Detergent-insoluble GPI-anchored proteins are apically sorted in fischer rat thyroid cells, but interference with cholesterol or sphingolipids differentially affects detergent insolubility and apical sorting. Mol Biol Cell 11:531–542

Lipschutz JH, Guo W, O'brien LE, Nguyen YH, Novick P, Mostov KE (2000) Exocyst is involved in cystogenesis and tubulogenesis and actsby modulating synthesis and delivery of basolateral plasma membrane and secretory proteins. Mol Biol Cell 11:4259–4275

Lisanti MP, Caras IW, Gilbert T, Hanzel D, Rodriguez-Boulan E (1990) Vectorial apical delivery and slow endocytosis of a glycolipid-anchored fusion protein in transfected MDCK cells. Proc Natl Acad Sci 87:7419–7423

Low SH, Chapin SJ, Weimbs T, Komuves LG, Bennett MK, Mostov KE (1996) Differential localization of syntaxin isoforms in polarized Madin-Darby canine kidney cells. Mol Biol Cell 7:2007–2018

Low SH, Chapin SJ, Wimmer C, Whiteheart SW, Komuves LG, Mostov KE, Weimbs T (1998a) The SNARE machinery is involved in apical plasma membrane trafficking in MDCK cells. J Cell Biol 141:1503–1513

Low SH, Roche PA, Anderson HA, Van Ijzendoorn SC, Zhang M, Mostov KE, Weimbs T (1998b) Targeting of SNAP-23 and SNAP-25 in polarized epithelial cells. J Biol Chem 273:3422–3430

Low SH, Miura M, Roche PA, Valdez AC, Mostov KE, Weimbs T (2000) Intracellular redirection of plasma membrane trafficking after loss of epithelial cell polarity. Mol Biol Cell 11:3045–3060

Lutcke A, Jansson S, Parton RG, Chavrier P, Valencia A, Huber LA, Lehtonen E, Zerial M (1993) Rab 17 a novel small GTPase, is specific for epithelial cells and is induced during cell polarization. J Cell Biol 121:553–564

Mah SP, Saueressig H, Goulding M, Kintner C, Dressler GR (2000) Kidney development in cadherin-6 mutants: delayed mesenchyme-to-epithelial conversion and loss of nephrons. Dev Biol 223:38–53

Makarova O, Roh MH, Liu CJ, Laurinec S, Margolis B (2003) Mammalian Crumbs3 is a small transmembrane protein linked to protein associated with Lin-7 (Pals1) Gene 302:21–29

Mandon B, Chou CL, Nielsen S, Knepper MA (1996) Syntaxin-4 is localized to the apical plasma membrane of rat renal collecting duct cells: possible role in aquaporin-2 trafficking. J Clin Invest 98:906–913

Mandon B, Nielsen S, Kishore BK, Knepper MA (1997) Expression of syntaxins in rat kidney. Am J Physiol 273:F718–F730

Maples CJ, Ruiz WG, Apodaca G (1997) Both microtubules and actin filaments are required for efficient postendocytotic traffic of the polymeric immunoglobulin receptor in polarized Madin-Darby canine kidney cells. J Biol Chem 272:6741–6751

Marples D, Barber B, Taylor A (1996) Effect of a dynein inhibitor on vasopressin action in toad urinary bladder. J Physiol 490:767–774

Marples D, Schroer TA, Ahrens N, Taylor A, Knepper MA, Nielsen S (1998) Dynein and dynactin colocalize with AQP2 water channels in intracellular vesicles from kidney collecting duct. Am J Physiol 274:F384–F394

Martin-Belmonte F, Puertollano R, Millan J, Alonso MA (2000) The MAL proteolipid is necessary for the overall apical delivery of membrane proteins in the polarized epithelial Madin-Darby canine kidney and fischer rat thyroid cell lines. Mol Biol Cell 11:2033–2045

Martin-Belmonte F, Martinez-Menarguez JA, Aranda JF, Ballesta J, De Marco MC, Alonso MA (2003) MAL regulates clathrin-mediated endocytosis at the apical surface of Madin-Darby canine kidney cells. J Cell Biol 163:155–164

Matern HT, Yeaman C, Nelson WJ, Scheller RH (2001) The Sec 6/8 complex in mammalian cells: characterization of mammalian Sec 3, subunit interactions, and expression of subunits in polarized cells. Proc Natl Acad Sci 98(17):9648–9653

Matlin KS, Simons K (1983) Reduced temperature prevents transfer of a membrane glycoprotein to the cell surface but does not prevent terminal glycosylation. Cell 34:233–243

Matter K, Mellman I (1994) Mechanisms of cell polarity: sorting and transport in epithelial cells. Curr Biol 6:545–554

Matter K, Hunziker W, Mellman I (1992) Basolateral sorting of LDL receptor in MDCK cells: the cytoplasmic domain contains two tyrosine-dependent targeting determinants. Cell 71:741–753

Matter K, Whitney JA, Yamamoto EM, Mellman I (1993) Common signals control low density lipoprotein receptor sorting in endosomes and the Golgi complex of MDCK cells. Cell 74:1053–1064

Matter K, Yamamoto EM, Mellman I (1994) Structural requirements and sequence motifs for polarized sorting and endocytosis of LDL and Fc receptors in MDCK cells. J Cell Biol 126:991–1004

Maximov A, Sudhof TC, Bezprozvanny I (1999) Association of neuronal calcium channels with modular adaptor proteins. J Biol Chem 274:24453–24456

Mays RW, Siemers KA, Fritz BA, Lowe AW, Van Meer G, Nelson WJ (1995) Hierarchy of mechanisms involved in generating Na/K-ATPase polarity in MDCK epithelial cells. J Cell Biol 130:1105–1115

Mcateer JA, Evan AP, Gardner KD (1987) Morphogenetic clonal growth of kidney epithelial cell line MDCK. Anat Rec 217:229–239

Mcbride HM, Rybin V, Murphy C, Giner A, Teasdale R, Zerial M (1999) Oligomeric complexes link Rab 5 effectors with NSF and drive membrane fusion via interactions between EEA1 and syntaxin 13. Cell 98:377–386

Mcneill H, Ozawa M, Kemler R, Nelson WJ (1990) Novel function of the cell adhesion molecule uvomorulin as an inducer of cell surface polarity. Cell 62:309–316

Melkonian KA, Chu T, Tortorella LB, Brown DA (1995) Characterization of proteins in detergent-resistant membrane complexes from Madin-Darby canine kidney epithelial cells. Biochemistry 34:16161–16170

Mellman I (1996) Endocytosis and molecular sorting. Annu Rev Cell Dev Biol 12:575–625

Mermall V, Post PL, Mooseker MS (1998) Unconventional myosins in cell movement, membrane traffic, and signal transduction. Science 279:527–533

Millan J, Puertollano R, Fan L, Rancano C, Alonso MA (1997) The MAL proteolipid is a component of the detergent-insoluble membrane subdomains of human T-lymphocytes. Biochem J 321:247–252

Miranda KC, Khromykh T, Christy P, Le TL, Gottardi CJ, Yap AS, Stow JL, Teasdale RD (2004) A dileucine motif targets E-cadherin to the basolateral cell surface in Madin-Darby canine kidney and LLC-PK1 epithelial cells. J Biol Chem 276:22565–22572

Misek DE, Bard E, Rodriguez-Boulan E (1984) Biogenesis of epithelial cell polarity: intracellular sorting and vectorial exocytosis of an apical plasma membrane glycoprotein. Cell 39:537–546

Mora R, Bonilha VL, Marmorstein A, Scherer PE, Brown D, Lisanti MP, Rodriguez-Boulan E (1999) Caveolin-2 localizes to the golgi complex but redistributes to plasma membrane, caveolae, and rafts when co-expressed with caveolin-1. J Biol Chem 274:25708–25717

Morris SM, Arden SD, Roberts RC, Kendrick-Jones J, Cooper JA, Luzio JP, Buss F (2004) Myosin VI binds to and localises with Dab2, potentially linking receptor-mediated endocytosis and the actin cytoskeleton. Traffic 3:331–341

Morrow JS, Cianci CD, Ardito T, Mann AS, Kashgarian M (1989) Ankyrin links fodrin to the alpha subunit of Na,K-ATPase in Madin-Darby canine kidney cells and in intact renal tubule cells. J Cell Biol 108:455–465

Moskalenko S, Henry DO, Rosse C, Mirey G, Camonis JH, White MA (2002) The exocyst is a Ral effector complex. Nat Cell Biol 4:66–72

Moskalenko S, Tong C, Rosse C, Camonis J, White MA (2003) Ral GTPases regulate exocyst assembly through dual subunit interactions. J Biol Chem (in press)

Mostov KE, Verges M, Altschuler Y (2000) Membrane traffic in polarized epithelial cells. Curr Opin Cell Biol 12:483–490

Moyer BD, Duhaime M, Shaw C, Denton J, Reynolds D, Karlson KH, Pfeiffer J, Wang S, Mickle JE, Milewski M, Cutting GR, Guggino WB, Li M, Stanton BA (2000) The PDZ-interacting domain of cystic fibrosis transmembrane conductance regulator is required for functional expression in the apical plasma membrane. J Biol Chem 275:27069–27074

Mukherjee S, Ghosh RN, Maxfield FR (1997) Endocytosis. Physiol Rev 77:759–803

Muller HA, Wieschaus E (1996) Amadillo, bazooka, and stardust are critical for early stages in formation of the zonula adherens and maintenance of the polarized blastoderm epithelium in Drosophila. J Cell Biol 134:149–163

Musch A, Cohen D, Rodriguez-Boulan E (1997) Myosin II is involved in the production of constitutive transport vesicles from the TGN J Cell Biol 138:291–301

Musch A, Cohen D, Kreitzer G, Rodriguez-Boulan E (2001) cdc42 regulates the exit of apical and basolateral proteins from the trans-Golgi network. EMBO J 20:2171–2179

Musch A, Cohen D, Yeaman C, Nelson WJ, Rodriguez-Boulan E, Brennwald PJ (2002) Mammalian homolog of Drosophila tumor suppressor lethal (2) giant larvae interacts with basolateral exocytic machinery in Madin-Darby canine kidney cells. Mol Biol Cell 13:158–168

Muth TR, Ahn J, Caplan MJ (1998) Identification of sorting determinants in the C-terminal cytoplasmic tails of the gamma-aminobutyric acid transporters GAT-2 and GAT-3. J Biol Chem 273:25616–25627

Nagai M, Meerloo T, Takeda T, Farquhar MG (2003) The adaptor protein ARH escorts megalin to and through endosomes. Mol Biol Cell 14:4984–4996

Nagai-Tamai Y, Mizuno K, Hirose T, Suzuki A, Ohno S (2002) Regulated protein–protein interaction between aPKC and PAR-3 plays an essential role in the polarization of epithelial cells. Genes Cells 7:1161–1171

Nakagawa S, Huibregtse JM (2000) Human scribble (Vartul) is targeted for ubiquitin-mediated degradation by the high-risk papillomavirus E6 proteins and the E6AP ubiquitin-protein ligase. Mol Cell Biol 20:8244–8253

Nathke IS, Hinck L, Swedlow JR, Papkoff J, Nelson WJ (1994) Defining interactions and distributions of cadherin and catenin complexes in polarized epithelial cells. J Cell Biol 125:1341–1352

Nelson WJ, Nusse R (2004) Convergence of Wnt, beta-catenin, and cadherin pathways. Science 303:1483–1487

Nelson WJ, Veshnock PJ (1987) Ankyrin binding to (Na+ + K+)ATPase and implications for the organization of membrane domains in polarized cells. Nature 328:533–536

Nelson WJ, Shore EM, Wang AZ, Hammerton RW (1990) Identification of a membrane-cytoskeletal complex containing the cell adhesion molecule uvomorulin (E-cadherin), ankyrin, and fodrin in Madin-Darby canine kidney epithelial cells. J Cell Biol 110:349–357

Nielsen S, Marples D, Birn H, Mohtashami M, Dalby NO, Trimble M, Knepper M (1995) Expression of VAMP-2-like protein in kidney collecting duct intracellular vesicles. Colocalization with Aquaporin-2 water channels. J Clin Invest 96:1834–1844

Noda Y, Okada Y, Saito N, Setou M, Xu Y, Zhang Z, Hirokawa N (2001) KIFC3 a microtubule minus end-directed motor for the apical transport of annexin XIIIb-associated Triton-insoluble membranes. J Cell Biol 155:77–88

Noren NK, Niessen CM, Gumbiner BM, Burridge K (2001) Cadherin engagement regulates Rho family GTPases. J Biol Chem 276:33305–33308

O'brien LE, Jou TS, Pollack AL, Zhang Q, Hansen SH, Yurchenco P, Mostov KE (2001) Rac1 orientates epithelial apical polarity through effects on basolateral laminin assembly. Nat Cell Biol 3:831–838

Odorizzi G, Trowbridge IS (1997) Structural requirements for basolateral sorting of the human transferrin receptor in the biosynthetic and endocytic pathways of Madin-Darby canine kidney cells. J Cell Biol 137:1255–1264

Odorizzi G, Pearse A, Domingo D, Trowbridge IS, Hopkins CR (1996) Apical and basolateral endosomes of MDCK cells are interconnected and contain a polarized sorting mechanism. J Cell Biol 135:139–152

Ohno H, Stewart J, Fournier MC, Bosshart H, Rhee I, Miyatake S, Gallusser A, Kirchhausen T, Bonifacino JS (1995) Interaction of tyrosine-based sorting signals with clathrin-associated proteins. Science 269:1872–1875

Ohno H, Fournier MC, Poy G, Bonifacino JS (1996) Structural determinants of interaction of tyrosine-based sorting signals with the adaptor medium chains. J Biol Chem 271:29009–29015

Ohno H, Aguilar RC, Yeh D, Taura D, Saito T, Bonifacino JS (1998) The medium subunits of adaptor complexes recognize distinct but overlapping sets of tyrosine-based sorting signals. J Biol Chem 273:25915–25921

Ohno H, Tomemori T, Nakatsu F, Okazaki Y, Aguilar RC, Foelsch H, Mellman I, Saito T, Shirasawa T, Bonifacino JS (1999) Mu1B a novel adaptor medium chain expressed in polarized epithelial cells. FEBS Lett 449:215–220

Ojakian GK, Schwimmer R (1988) The polarized distribution of an apical cell surface glycoprotein is maintained by interactions with the cytoskeleton of Madin-Darby canine kidney cells. J Cell Biol 107:2377–2387

Okamoto M, Sudhof TC (1997) Mints, Munc18-interacting proteins in synaptic vesicle exocytosis. J Biol Chem 272:31459–31464

Olsen O, Liu H, Wade JB, Merot J, Welling PA (2002) Basolateral membrane expression of the Kir 2.3 channel is coordinated by PDZ interaction with Lin-7/CASK complex. Am J Physiol Cell Physiol 282:C183–C195

Orzech E, Cohen S, Weiss A, Aroeti B (2000) Interactions between the exocytic and endocytic pathways in polarized Madin-Darby canine kidney cells. J Biol Chem 275:15207–15219

Owen DJ, Evans PR (1998) A structural explanation for the recognition of tyrosine-based endocytotic signals. Science 282:1327–1332

Palade G (1975) Intracellular aspects of the process of protein synthesis. Science 189:347–358

Parton RG, Prydz K, Bomsel M, Simons K, Griffiths G (1989) Meeting of the apical and basolateral endocytic pathways of the Madin-Darby canine kidney cell in late endosomes. J Cell Biol 109:3259–3272

Pelham HR (2001) SNAREs and the specificity of membrane fusion. Trends Cell Biol 11:99–101

Perego C, Bulbarelli A, Longhi R, Caimi M, Villa A, Caplan MJ (1997) Sorting of two polytopic proteins, the gamma-aminobutyric acid and betaine transporters, in polarized epithelial cells. J Biol Chem 272:6584–6592

Perego C, Vanoni C, Villa A, Longhi R, Kaech SM, Frohli E, Hajnal A, Kim SK, Pietrini G (1999) PDZ-mediated interactions retain the epithelial GABA transporter on the basolateral surface of polarized epithelial cells. EMBO J 18:2384–2393

Pfeiffer S, Fuller SD, Simons K (1985) Intracellular sorting and basolateral appearance of the G protein of vesicular stomatitis virus in Madin-Darby canine kidney cells. J Cell Biol 101:470–476

Polishchuk R, Pentima AD, Lippincott-Schwartz J (2004) Delivery of raft-associated, GPI-anchored proteins to the apical surface of polarized MDCK cells by a transcytotic pathway. Nat Cell Biol 6:297–307

Potter BA, Ihrke G, Bruns JR, Weixel KM, Weisz OA (2004) Specific N-glycans direct apical delivery of transmembrane, but not soluble or glycosylphosphatidylinositol-anchored forms of endolyn in Madin-Darby canine kidney cells. Mol Biol Cell 15:1407–1416

Powell SK, Cunningham BA, Edelman GM, Rodriguez-Boulan E (1991) Targeting of transmembrane and GPI-anchored forms of N-CAM to opposite domains of a polarized epithelial cell. Nature 353:76–77

Prekeris R, Klumperman J, Scheller RH (2000) A Rab 11/Rip11 protein complex regulates apical membrane trafficking via recycling endosomes. Mol Cell 6:1437–1438

Puertollano R, Martin-Belmonte F, Millan J, De Marco MC, Albar JP, Kremer L, Alonso MA (1999) The MAL proteolipid is necessary for normal apical transport and accurate sorting of the influenza virus hemagglutinin in Madin-Darby canine kidney cells. J Cell Biol 145:141–151

Puertollano R, Aguilar RC, Gorshkova I, Crouch RJ, Bonifacino JS (2001) Sorting of mannose 6-phosphate receptors mediated by the GGAs. Science 292:1712–1716

Quinones B, Riento K, Olkkonen VM, Hardy S, Bennett MK (1999) Syntaxin 2 splice variants exhibit differential expression patterns, biochemical properties and subcellular localizations. J Cell Sci 112:4291–4304

Rajasekaran AK, Hojo M, Huima T, Rodriguez-Boulan E (1996) Catenins and zonula occludens-1 form a complex during early stages in the assembly of tight junctions. J Cell Biol 132:451–463

Rapoport I, Chen YC, Cupers P, Shoelson SE, Kirchhausen T (1998) Dileucine-based sorting signals bind to the beta chain of AP-1 at a site distinct and regulated differently from the tyrosine-based motif-binding site. EMBO J 17:2148–2155

Reich V, Mostov K, Aroeti B (1996) The basolateral sorting signal of the polymeric immunoglobulin receptor contains two functional domains. J Cell Sci 109:2133–2139

Reuver SM, Garner CC (1998) E-cadherin mediated cell adhesion recruits SAP97 into the cortical cytoskeleton. J Cell Sci 111:1071–1080

Riento K, Galli T, Jansson S, Ehnholm C, Lehtonen E, Olkkonen VM (1998) Interaction of Munc-18–2 with syntaxin 3 controls the association of apical SNAREs in epithelial cells. J Cell Sci 111:2681–2688

Rindler MJ, Ivanov IE, Plesken H, Rodriguez-Boulan E, Sabatini DD (1984) Viral glycoproteins destined for apical or basolateral plasma membrane domains traverse the same Golgi apparatus during their intracellular transport in doubly infected Madin-Darby canine kidney cells. J Cell Biol 98:1304–1319

Rindler MJ, Ivanov IE, Sabatini DD (1987) Microtubule-acting drugs lead to the nonpolarized delivery of the influenza hemagglutinin to the cell surface of polarized Madin-Darby canine kidney cells. J Cell Biol 104:231–241

Robinson MS, Bonifacino JS (2001) Adaptor-related proteins. Curr Opin Cell Biol 13:444–453

Rodriguez OC, Cheney RE (2002) Human myosin-Vc is a novel class V myosin expressed in epithelial cells. J Cell Sci 115:991–1004

Rodriguez-Boulan E, Gonzalez A (1999) Glycans in post-Golgi apical targeting: sorting signals or structural props? Trends Cell Biol 9:291–224

Rodriguez-Boulan E, Nelson WJ (1989) Morphogenesis of the polarized epithelial cell phenotype. Science 245:718–725

Rodriguez-Boulan ER, Pendergast M (1980) Polarized distribution of viral envelope proteins in the plasma membrane of infected epithelial cells. Cell 20:45–54

Rodriguez-Boulan ER, Sabatini DD (1978) Asymmetric budding of viruses in epithelial monolayers: a model system for study of epithelial polarity. Proc Natl Acad Sci 75:5071–5075

Rodriguez-Boulan E, Paskiet KT, Sabatini DD (1983) Assembly of enveloped viruses in Madin-Darby canine kidney cells: polarized budding from single attached cells and from clusters of cells in suspension. J Cell Biol 96:866–874

Rodriguez-Boulan E, Paskiet KT, Salas PJ, Bard E (1984) Intracellular transport of influenza virus hemagglutinin to the apical surface of Madin-Darby canine kidney cells. J Cell Biol 98:308–319

Roh MH, Liu CJ, Laurinec S, Margolis B (2002a) The carboxyl terminus of zona occludens-3 binds and recruits a mammalian homologue of discs lost to tight junctions. J Biol Chem 277:27501–27509

Roh MH, Makarova O, Liu CJ, Shin K, Lee S, Laurinec S, Goyal M, Wiggins R, Margolis B (2002b) The Maguk protein, Pals1, functions as an adapter, linking mammalian homologues of Crumbs and Discs Lost. J Cell Biol 157:161–172

Rojas R, Apodaca G (2002) Immunoglobulin transport across polarized epithelial cells. Nat Rev Mol Cell Biol 3:944–955

Roper K, Corbeil D, Huttner WB (2000) Retention of prominin in microvilli reveals distinct cholesterol-based lipid micro-domains in the apical plasma membrane. Nat Cell Biol 2:582–592

Rossetto O, Gorza L, Schiavo G, Schiavo N, Scheller RH, Montecucco C (1996) VAMP/synaptobrevin isoforms 1 and 2 are widely and differentially expressed in nonneuronal tissues. J Cell Biol 132:167–179

Rothman JE (1994) Mechanisms of intracellular protein transport. Nature 372:55–63

Roush DL, Gottardi CJ, Naim HY, Roth MG, Caplan MJ (1998) Tyrosine-based membrane protein sorting signals are differentially interpreted by polarized Madin-Darby canine kidney and LLC-PK1 epithelial cells. J Biol Chem 273:26862–26869

Rowe J, Calegari F, Taverna E, Longhi R, Rosa P (2001) Syntaxin 1A is delivered to the apical and basolateral domains of epithelial cells: the role of munc-18 proteins. J Cell Sci 114:3323–3332

Sabolic I, Katsura T, Verbavatz JM, Brown D (1995) The AQP2 water channel: effect of vasopressin treatment, microtubule disruption, and distribution in neonatal rats. J Mol Biol 143:165–175

Sabolic I, Herak-Kramberger CM, Ljubojevic M, Biemesderfer D, Brown D (2002) NHE3 and NHERF are targeted to the basolateral membrane in proximal tubules of colchicine-treated rats. Kidney Int 61:1351–1364

Saelman EU, Keely PJ, Santoro SA (1995) Loss of MDCK cell alpha 2 beta 1 integrin expression results in reduced cyst formation, failure of hepatocyte growth factor/scatter factor-induced branching morphogenesis, and increased apoptosis. J Cell Sci 108:3531–3540

Salas PJ, Vega-Salas DE, Hochman J, Rodriguez-Boulan E, Edidin M (1988) Selective anchoring in the specific plasma membrane domain: a role in epithelial cell polarity. J Cell Biol 107:2363–2376

Sanchez-Pulido L, Martin-Belmonte F, Valencia A, Alonso MA (2002) MARVEL: a conserved domain involved in membrane apposition events. Trends Biochem Sci 27:599–601

Santoni MJ, Pontarotti P, Birnbaum D, Borg JP (2002) The LAP family: a phylogenetic point of view. Trends Genet 18:494–497

Saraste J, Kuismanen E (1984) Pre- and post-Golgi vacuoles operate in the transport of Semliki Forest virus membrane glycoproteins to the cell surface. Cell 38:535–549

Sargiacomo M, Sudol M, Tang Z, Lisanti MP (1993) Signal transducing molecules and glycosyl-phosphatidylinositol-linked proteins form a caveolin-rich insoluble complex in MDCK cells. J Cell Biol 122:789–807

Saunders C, Limbird LE (1997) Disruption of microtubules reveals two independent apical targeting mechanisms for G-protein-coupled receptors in polarized renal epithelial cells. J Biol Chem 272:19035–19045

Scheiffele P, Peranen J, Simons K (1995) N-glycans as apical sorting signals in epithelial cells. Nature 378:96–98

Schimmoller F, Simon I, Pfeffer SR (1998) Rab GTPases, directors of vesicle docking. J Biol Chem 273:22161–22164

Schwartz GJ, Tsuruoka S, Vijayakumar S, Petrovic S, Mian A, Al-Awqati Q (2002) Acid incubation reverses the polarity of intercalated cell transporters, an effect mediated by hensin. J Clin Invest 109:89–99

Schwimmer R, Ojakian GK (1995) The alpha 2 beta 1 integrin regulates collagen-mediated MDCK epithelial membrane remodeling and tubule formation. J Cell Sci 108:2487–2498

Scott DB, Blanpied TA, Swanson GT, Zhang C, Ehlers MD (2001) An NMDA receptor ER retention signal regulated by phosphorylation and alternative splicing. J Neurosci 21:3063–3072

Scott RO, Thelin WR, Milgram SL (2002) A novel PDZ protein regulates the activity of guanylyl cyclase C, the heat-stable enterotoxin receptor. J Biol Chem 277:22934–22941

Setou M, Nakagawa T, Seog DH, Hirokawa N (2000) Kinesin superfamily motor protein KIF17 and mLin-10 in NMDA receptor-containing vesicle transport. Science 288:1796–1802

Sheff DR, Daro EA, Hull M, Mellman I (1999) The receptor recycling pathway contains two distinct populations of early endosomes with different sorting functions. J Cell Biol 145:123–139

Sheff DR, Kroschewski R, Mellman I (2002) Actin dependence of polarized receptor recycling in Madin-Darby canine kidney cell endosomes. Mol Biol Cell 13:262–275

Sheikh H, Isacke CM (1996) A di-hydrophobic Leu-Val motif regulates the basolateral localization of CD44 in polarized Madin-Darby canine kidney epithelial cells. J Biol Chem 271:12185–12190

Sheng M, Sala C (2001) PDZ domains and the organization of supramolecular complexes. Annu Rev Neurosci 24:1–29

Shenolikar S, Weinman EJ (2001) NHERF: targeting and trafficking membrane proteins. Am J Physiol Renal Physiol 280:F389–F395

Shenolikar S, Voltz JW, Minkoff CM, Wade JB, Weinman EJ (2002) Targeted disruption of the mouse NHERF-1 gene promotes internalization of proximal tubule sodium-phosphate cotransporter type IIa and renal phosphate wasting. Proc Natl Acad Sci 99:11470–11475

Short DB, Trotter KW, Reczek D, Kreda SM, Bretscher A, Boucher RC, Stutts MJ, Milgram SL (1998) An apical PDZ protein anchors the cystic fibrosis transmembrane conductance regulator to the cytoskeleton. J Biol Chem 273:19797–19801

Shurety W, Stewart NL, Stow JL (1998) Fluid-phase markers in the basolateral endocytic pathway accumulate in response to the actin assembly-promoting drug Jasplakinolide. Mol Biol Cell 9:957–975

Simmen T, Nobile M, Bonifacino JS, Hunziker W (1999) Basolateral sorting of furin in MDCK cells requires a phenylalanine-isoleucine motif together with an acidic amino acid cluster. Mol Cell Biol 19:3136–3144

Simmen T, Honing S, Icking A, Tikkanen R, Hunziker W (2002) AP-4 binds basolateral signals and participates in basolateral sorting in epithelial MDCK cells. Nat Cell Biol 4:154–159

Simon JP, Shen TH, Ivanov IE, Gravotta D, Morimoto T, Adesnik M, Sabatini DD (1998) Coatomer, but not P200/myosin II, is required for the in vitro formation of trans-Golgi network-derived vesicles containing the envelope glycoprotein of vesicular stomatitis virus. Proc Natl Acad Sci 95:1073–1080

Simons K, Van Meer G (1988) Lipid sorting in epithelial cells. Biochemistry 27:6197–6202

Simonsen A, Lippe R, Christoforidis S, Gaullier JM, Brech A, Callaghan J, Toh BH, Murphy C, Zerial M, Stenmark H (1998) EEA1 links PI(3)K function to Rab 5 regulation of endosome fusion. Nature 394:494–498

Simonsen A, Gaullier JM, D'arrigo A, Stenmark H (1999a) The Rab 5 effector EEA1 interacts directly with syntaxin-6. J Biol Chem 274:28857–28860

Simonsen A, Pedersen KW, Nordeng TW, Von Der Lippe A, Stang E, Long EO, Bakke O (1999b) Polarized transport of MHC class II molecules in Madin-Darby canine kidney cells is directed by a leucine-based signal in the cytoplasmic tail of the beta-chain. J Immunol 163:2540–2548

Simske JS, Kaech SM, Harp SA, Kim SK (1996) LET-23 receptor localization by the cell junction protein LIN-7 during C. elegans vulval induction. Cell 85:195–204

Sollner T, Bennett MK, Whiteheart SW, Scheller RH, Rothman JE (1993a) A protein assembly-disassembly pathway in vitro that may correspond to sequential steps of synaptic vesicle docking, activation, and fusion. Cell 75:409–418

Sollner T, Whiteheart SW, Brunner M, Erdjument-Bromage H, Geromanos S, Tempst P, Rothman JE (1993b) SNAP receptors implicated in vesicle targeting and fusion. Nature 362:318–324

Songyang Z, Fanning AS, Fu C, Xu J, Marfatia SM, Chishti AH, Chan AC, Anderson JM, Cantley LC (1997) Recognition of unique carboxyl-terminal motifs by distinct PDZ domains. Science 275:73–77

Standley S, Roche KW, Mccallum J, Sans N, Wenthold RJ (2000) PDZ domain suppression of an ER retention signal in NMDA receptor NR1 splice variants. Neuron 28:887–898

Steegmaier M, Lee KC, Prekeris R, Scheller RH (2000) SNARE protein trafficking in polarized MDCK cells. Traffic 1:553–560

Stewart DB, Barth AI, Nelson WJ (2000) Differential regulation of endogenous cadherin expression in Madin-Darby canine kidney cells by cell–cell adhesion and activation of beta-catenin signaling. J Biol Chem 275:20707–20716

Stow JL (1995) Regulation of vesicular transport by GTP-binding proteins. Curr Opin Nephrol Hypertens 4:421–425

Straight SW, Karnak D, Borg JP, Kamberov E, Dare H, Margolis B, Wade JB (2000) mLin-7 is localized to the basolateral surface of renal epithelia via its NH(2) terminus. Am J Physiol Renal Physiol 278:F464–F475

Straight SW, Chen L, Karnak D, Margolis B (2001) Interaction with mLin-7 alters the targeting of endocytosed transmembrane proteins in mammalian epithelial cells. Mol Biol Cell 12:1329–1340

Sugihara K, Asano S, Tanaka K, Iwamatsu A, Okawa K, Ohta Y (2002) The exocyst complex binds the small GTPase RalA to mediate filopodia formation. Nat Cell Biol 4:73–78

Sugimoto H, Sugahara M, Folsch H, Koide Y, Nakatsu F, Tanaka N, Nishimura T, Furukawa M, Mullins C, Nakamura N, Mellman I, Ohno H (2002) Differential recognition of tyrosine-based basolateral signals by AP-1B subunit mu1B in polarized epithelial cells. Mol Biol Cell 13:2374–2382

Sutton RB, Fasshauer D, Jahn R, Brunger AT (1998) Crystal structure of a SNARE complex involved in synaptic exocytosis at 2.4 A resolution. Nature 395:347–353

Suzuki A, Yamanaka T, Hirose T, Manabe N, Mizuno K, Shimizu M, Akimoto K, Izumi Y, Ohnishi T, Ohno S (2001) Atypical protein kinase C is involved in the evolutionarily conserved par protein complex and plays a critical role in establishing epithelia-specific junctional structures. J Cell Biol 152:1183–1196

Tachibana K, Nakanishi H, Mandai K, Ozaki K, Ikeda W, Yamamoto Y, Nagafuchi A, Tsukita S, Takai Y (2000) Two cell adhesion molecules, nectin and cadherin, interact through their cytoplasmic domain-associated proteins. J Cell Biol 150:1161–1176

Tai AW, Chuang JZ, Sung CH (2001) Cytoplasmic dynein regulation by subunit heterogeneity and its role in apical transport. J Cell Biol 153:1499–1509

Takaishi K, Sasaki T, Kotani H, Nishioka H, Takai Y (1997) Regulation of cell–cell adhesion by rac and rho small G proteins in MDCK cells. J Cell Biol 139:1047–1059

Takeda T, Yamazaki H, Farquhar MG (2003) Identification of an apical sorting determinant in the cytoplasmic tail of megalin. Am J Physiol Cell Physiol 284:C1105–1113

Takekuni K, Ikeda W, Fujito T, Morimoto K, Takeuchi M, Monden M, Takai Y (2003) Direct binding of cell polarity protein PAR-3 to cell–cell adhesion molecule nectin at neuroepithelial cells of developing mouse. J Biol Chem 278:5497–5500

Tall RD, Alonso MA, Roth MG (2004) Features of influenza HA required for apical sorting differ from those required for association with DRMs or MAL. Traffic 4:838–849

Tanentzapf G, Tepass U (2003) Interactions between the crumbs, lethal giant larvae and bazooka pathways in epithelial polarization. Nat Cell Biol 5:46–52

Tepass U, Knust E (1993) Crumbs and stardust act in a genetic pathway that controls the organization of epithelia in Drosophila melanogaster. Dev Biol 159:311–326

Tepass U, Theres C, Knust E (1990) crumbs encodes an EGF-like protein expressed on apical membranes of Drosophila epithelial cells and required for organization of epithelia. Cell 61:787–799

Tepass U, Tanentzapf G, Ward R, Fehon R (2003) Epithelial cell polarity and cell junctions in Drosophila. Annu Rev Genet 35:747–784

Terbush DR, Maurice T, Roth D, Novick P (1996) The Exocyst is a multiprotein complex required for exocytosis in Saccharomyces cerevisiae. EMBO J 15:6483–6994

Thomas DC, Roth MG (1994) The basolateral targeting signal in the cytoplasmic domain of glycoprotein G from vesicular stomatitis virus resembles a variety of intracellular targeting motifs related by primary sequence but having diverse targeting activities. J Biol Chem 269:15732–15739

Thomas DC, Brewer CB, Roth MG (1993) Vesicular stomatitis virus glycoprotein contains a dominant cytoplasmic basolateral sorting signal critically dependent upon a tyrosine. J Biol Chem 268:3313–3320

Tojo A, Bredt DS, Wilcox CS (1999) Distribution of postsynaptic density proteins in rat kidney: relationship to neuronal nitric oxide synthase. Kidney Int 55:1384–1394

Traub LM, Apodaca G (2003) AP-1B: polarized sorting at the endosome. Nat Cell Biol 5:1045–1047

Tuma PL, Hubbard AL (2003) Transcytosis: crossing cellular barriers. Physiol Rev 83:871–932

Vale RD (2003) The molecular motor toolbox for intracellular transport. Cell 112:467–480

Van Adelsberg J, Edwards JC, Takito J, Kiss B, Al-Awqati Q (1994) An induced extracellular matrix protein reverses the polarity of band 3 in intercalated epithelial cells. Cell 76:1053–1061

Van Ijzendoorn SCD, Maier O, Van Der Wouden JM, Hoekstra D (2000) The subapical compartment and its role in intracellular trafficking and cell polarity. J Cell Physiol 184:151–160

Van Ijzendoorn SCD, Tuvim MJ, Weimbs T, Dickey BF, Mostov KE (2002) Direct interaction between Rab 3b and the polymeric immunoglobulin receptor controls ligand-stimulated transcytosis in epithelial cells. Dev Cell 2:219–228

Vega-Salas DE, Salas PJ, Gundersen D, Rodriguez-Boulan E (1987a) Formation of the apical pole of epithelial (Madin-Darby canine kidney) cells: polarity of an apical protein is independent of tight junctions while segregation of a basolateral marker requires cell–cell interactions. J Cell Biol 104:905–916

Vega-Salas DE, Salas PJ, Rodriguez-Boulan E (1987b) Modulation of the expression of an apical plasma membrane protein of Madin-Darby canine kidney epithelial cells: cell–cell interactions control the appearance of a novel intracellular storage compartment. J Cell Biol 104:1249–1259

Vega-Salas DE, Salas PJ, Rodriguez-Boulan E (1988) Exocytosis of vacuolar apical compartment (VAC): a cell–cell contact controlled mechanism for the establishment of the apical plasma membrane domain in epithelial cells. J Cell Biol 107:1717–1728

Verkade P, Simons K (1997) Robert Feulgen Lecture 1997. Lipid microdomains and membrane trafficking in mammalian cells. Histochem Cell Biol 108:211–220

Vogel U, Sandvig K, Van Deurs B (1998) Expression of caveolin-1 and polarized formation of invaginated caveolae in Caco-2 and MDCK II cells. J Cell Sci 111:825–832

Von Bonsdorff CH, Fuller SD, Simons K (1985) Apical and basolateral endocytosis in Madin-Darby canine kidney (MDCK) cells grown on nitrocellulose filters. EMBO J 4:2781–2792

Wade JB, Welling PA, Donowitz M, Shenolikar S, Weinman EJ (2001) Differential renal distribution of NHERF isoforms and their colocalization with NHE3, ezrin, and ROMK. Am J Physiol Cell Physiol 280:C192–C198

Wagner M, Morgans C, Koch-Brandt C (1995) The oligosaccharides have an essential but indirect role in sorting gp80 (clusterin, TRPM-2) to the apical surface of MDCK cells. Eur J Cell Biol 67:84–88

Wan J, Taub ME, Shah D, Shen WC (1992) Brefeldin A enhances receptor-mediated transcytosis of transferrin in filter-grown Madin-Darby canine kidney cells. J Biol Chem 267:13446–13450

Wandinger-Ness A, Bennett MK, Antony C, Simons K (1990) Distinct transport vesicles mediate the delivery of plasma membrane proteins to the apical and basolateral domains of MDCK cells. J Cell Biol 111:987–1000

Wang AZ, Ojakian GK, Nelson WJ (1990a) Steps in the morphogenesis of a polarized epithelium. I. Uncoupling the roles of cell–cell and cell–substratum contact in establishing plasma membrane polarity in multicellular epithelial (MDCK) cysts. J Cell Sci 95:137–151

Wang AZ, Ojakian GK, Nelson WJ (1990b) Steps in the morphogenesis of a polarized epithelium. II. Disassembly and assembly of plasma membrane domains during reversal of epithelial cell polarity in multicellular epithelial (MDCK) cysts. J Cell Sci 95:153–165

Wang E, Brown PS, Aroeti B, Chapin SJ, Mostov KE, Dunn KW (2000) Apical and basolateral endocytic pathways of MDCK cells meet in acidic common endosomes distinct from a nearly neutral apical recycling endosome. Traffic 1:480–493

Wang E, Pennington JG, Goldenring JR, Hunziker W, Dunn KW (2002) Brefeldin A rapidly disrupts plasma membrane polarity by blocking polar sorting in common endosomes of MDCK cells. J Cell Sci 114:3309–3321

Wang S, Raab RW, Schatz PJ, Guggino WB, Li M (1998) Peptide binding consensus of the NHE-RF-PDZ1 domain matches the C-terminal sequence of CFTR. FEBS Lett 427:103–108

Wang S, Yue H, Derin RB, Guggino WB, Li M (2001) Accessory protein facilitated CFTR-CFTR interaction a molecular mechanism to potentiate the chloride channel activity. Cell 103:169–179

Wang X, Kumar R, Navarre J, Casanova JE, Goldenring JR (2000) Regulation of vesicle trafficking in Madin-Darby canine kidney cells by Rab 11a and Rab 25. J Biol Chem 275:29138–29146

Waters MG, Pfeffer SR (1999) Membrane tethering in intracellular transport. Curr Opin Cell Biol 11:453–459

Wehrle-Haller B, Imhof BA (2001) Stem cell factor presentation to c-Kit. Identification of a basolateral targeting domain. J Biol Chem 276:12667–12674

Weinman EJ, Boddeti A, Cunningham R, Akom M, Wang F, Wang Y, Liu J, Steplock D, Shenolikar S, Wade JB (2003) NHERF-1 is required for renal adaptation to a low-phosphate diet. Am J Physiol Renal Physiol 285:F1225–F1232

Wells AL, Lin AW, Chen LQ, Safer D, Cain SM, Hasson T, Carragher BO, Milligan RA, Sweeney HL (1999) Myosin VI is an actin-based motor that moves backwards. Nature 401:505–508

Wilson JM, De Hoop M, Zorzi N, Toh BH, Dotti CG, Parton RG (2000) EEA1 a tethering protein of the early sorting endosome, shows a polarized distribution in hippocampal neurons, epithelial cells, and fibroblasts. Mol Biol Cell 11:2657–2671

Wodarz A, Hinz U, Engelbert M, Knust E (1995) Expression of crumbs confers apical character on plasma membrane domains of ectodermal epithelia of Drosophila. Cell 82:67–76

Wu H, Reuver SM, Kuhlendahl S, Chung WJ, Garner CC (1998) Subcellular targeting and cytoskeletal attachment of SAP97 to the epithelial lateral membrane. J Cell Sci 111:2365-276

Xu XZ, Choudhury A, Li X, Montell C (1998) Coordination of an array of signaling proteins through homo- and heteromeric interactions between PDZ domains and target proteins. J Cell Biol 142:545–555

Yamanaka T, Horikoshi Y, Suzuki A, Sugiyama Y, Kitamura K, Maniwa R, Nagai Y, Yamashita A, Hirose T, Ishikawa H, Ohno S (2001) PAR-6 regulates aPKC activity in a novel way and mediates cell–cell contact-induced formation of the epithelial junctional complex. Genes Cells 6:721–731

Yasumi M, Shimizu K, Honda T, Takeuchi M, Takai Y (2003) Role of each immunoglobulin-like loop of nectin for its cell–cell adhesion activity. Biochem Biophys Res Commun 302:61–66

Yeaman C, Le Gall AH, Baldwin AN, Monlauzeur L, Le Bivic A, Rodriguez-Boulan E (1997) The O-glycosylated stalk domain is required for apical sorting of neurotrophin receptors in polarized MDCK cells. J Cell Biol 139:929–940

Yeaman C, Grindstaff KK, Nelson WJ (1999) New perspectives on mechanisms involved in generating epithelial cell polarity. Physiol Rev 79:73–98

Yeaman C, Grindstaff KK, Wright JR, Nelson WJ (2001) Sec 6/8 complexes on trans-Golgi network and plasma membrane regulate late stages of exocytosis in mammalian cells. J Cell Biol 155:593–604

Yokoyama S, Tachibana K, Nakanishi H, Yamamoto Y, Irie K, Mandai K, Nagafuchi A, Monden M, Takai Y (2001) alpha-catenin-independent recruitment of ZO-1 to nectin-based cell–cell adhesion sites through afadin. Mol Biol Cell 12:1595–1609

Zacchetti D, Peranen J, Murata M, Fiedler K, Simons K (1995) VIP17/MAL a proteolipid in apical transport vesicles. FEBS Lett 377:465–469

Zacchi P, Stenmark H, Parton RG, Orioli D, Lim F, Giner A, Mellman I, Zerial M, Murphy C (1998) Rab 17 regulates membrane trafficking through apical recycling endosomes in polarized epithelial cells. J Cell Biol 140:1039–1053

Zhu Y, Doray B, Poussu A, Lehto VP, Kornfeld S (2001) Binding of GGA2 to the lysosomal enzyme sorting motif of the mannose 6-phosphate receptor. Science 292:1716–1718

Zurzolo C, Lisanti MP, Caras IW, Nitsch L, Rodriguez-Boulan E (1993) Glycosylphosphatidylinositol-anchored proteins are preferentially targeted to the basolateral surface in Fischer rat thyroid epithelial cells. J Cell Biol 121:1031–1039

Rev Physiol Biochem Pharmacol (2005) 153:101–111
DOI 10.1007/s10254-004-0024-6

M. A. Jakupec · P. Unfried · B. K. Keppler

Pharmacological properties of cerium compounds

Published online: 30 November 2004

Abstract Cerium is a member of the lanthanide series or rare earth elements which exert diverse biological effects mainly by their resemblance to calcium. This similarity, which is particularly characteristic for the lighter members of the lanthanide series, enables these elements to replace calcium in biomolecules without necessarily substituting for it functionally. While the inhibitory effects on calcium-dependent physiological processes (such as those involved in the blood clotting cascade as well as in neuronal and muscular functions) are well-known, their relevance for the pharmacological properties of cerium are less clear. Historically, cerium oxalate was used as an antiemetic, especially in vomiting of pregnancy and kinetoses, although its mechanism of action has never been clarified. At present, cerium nitrate is available as an adjunct to silver sulfadiazine cream for the topical treatment of extensive burns not amenable to early wound excision. Apart from direct antiseptic effects, cerium helps to prevent postburn sepsis and systemic inflammatory response by fixing burn toxins. The antineoplastic potential of cerium compounds, which had fallen into oblivion, is currently being re-explored in experimental settings, though the mechanistic basis remains to be elucidated.

Introductory remark

Cerium is a member of the lanthanides, lanthanoids, or rare earth elements. The definitions of these terms differ somewhat: while lanthanides (abbreviated Ln) is the historical name for the "d-element" lanthanum (La, atomic number: 57) and the following 14 "f-elements" from cerium (Ce, atomic number: 58) to lutetium (Lu, atomic number: 71), the term lanthanoids, created by the IUPAC in 1963, comprise only the "f-elements" from Ce to Lu, excluding La. Rare earth elements is the historical term still in use for the lanthanides plus yttrium (Y) and scandium (Sc), reflecting the fact that these two elements are associated with the lanthanides in their natural occurrence.

M. A. Jakupec · P. Unfried · B. K. Keppler (✉)
Institute of Inorganic Chemistry, University of Vienna,
Waehringer Strasse 42, 1090 Vienna, Austria
e-mail: bernhard.keppler@univie.ac.at · Tel.: +43-1-427752600 · Fax: +43-1-42772680

The discovery of the rare earth elements began in the year 1787, when Arrhenius identified a grayish-black and very heavy mineral which he named "ytterbite." This mineral turned out to be a real treasure, containing a lot of previously unknown elements and setting off an exciting quest for new elements (Kleber 1961; Moeller 1963, 1973; Trifonov 1963; Eyring 1964). First, subgroups were isolated, called "ceria," "terbia," "yttria," and "ytterbia," corresponding to mixtures of the light ("cerit-earth elements"), middle ("terbin-earth elements"), heavy ("ytter-earth elements"), and very heavy earths ("ytterbin-earth elements"). The similar chemical behavior of the components of these fractions made the first analyzers consider these mixtures as pure elements. In 1803, cerium became the first of the lanthanides to be isolated with high purity. Still, for most early applications purity was not important, and therefore one must be aware that the term cerium when used in older literature (up to the early twentieth century) does not always mean pure cerium.

Chemical and biochemical aspects

To understand the chemical behavior of cerium and its unique position among the lanthanides, it is necessary to discuss some aspects of the chemistry of this family of very similar elements. In physiological solutions, most of the lanthanides are only stable in the trivalent form, the only exceptions being Ce(IV) and Eu(II) (Asprey and Cunningham 1960). The ionic radii of trivalent ions (in complexes with coordination number 6) steadily decrease with increasing atomic number within the small range from 1.03 Å to 0.86 Å, a phenomenon known as the lanthanide contraction. The small differences in ionic hardness result in slight differences of their distribution coefficients and the hydrolytic sensibility of their salts. Both effects are relevant for the purification of the individual elements. Ce(IV) (ionic radius: 0.89 Å) shows a hydrolytic behavior like a hypothetical "super-lutetium." Therefore, aqueous solutions of Ce(IV) salts (without complexation) are stable only if acidified to $pH<3$. Otherwise, hydrolysis takes place, together with formation of some turbidity or gel at higher concentrations. Since weakly bound ligands like biological buffers do not prevent hydrolysis, Ce(IV) salts have no biological relevance.

The preferred coordination number of all lanthanides is 8–9, but complexes from 6–12 could be prepared as well. In aqueous solution, the hydration shell includes 9–12 water molecules—a controversy about these values is still going on. The ionic character of complexes of trivalent lanthanide ions is very high, with a typical order in strength of bonding to $O<N<S$. The interaction with nitrogen in complexes such as those with o-phenanthroline (Sinha 1966) is very weak, and ligands containing only N-donor atoms do not complex with rare earth ions under physiological conditions. The predominant donor in aqueous solutions is the oxygen atom. Sugars, nucleosides, and ligands with carboxyl groups are preferential complexing groups, and chelating is the predominant form of complexation (Komiyama and Sumaoka 2001).

The redox potential ϵ^{o} of the system Ce^{3+}/Ce^{4+} is 1.70 V (in 1 M $HClO_4$). In comparison with the standard potential of the redox system H_2O/O_2 (1.299 V) and the possible formation of Ce(III) and O_2, the meta-stable character of aqueous Ce(IV) solutions can be understood. Ce(IV) is a powerful oxidant without biological relevance (Nash and Sullivan 1991), but it has been studied as a catalyst of phosphate ester and nucleic acid hydrolysis (Sumaoka et al. 1998; Takarada et al. 2000; Shigekawa et al. 1999). On the other hand, Ce(III) is very resistant to oxidation and reacts only with very strong oxidants.

Calcium in the bivalent state (ionic radius: 1.00 Å) is very similar to trivalent cerium (ionic radius: 1.01 Å) in terms of size, bonding, and preferences to donor atoms. Hence, it is not surprising that trivalent cerium behaves chemically similar to calcium. One of the most characteristic differences between Ln(III) and Ca(II) is the higher charge-to-volume-ratio of the former, resulting in a stronger binding to water molecules, an increased stability of analogous complexes, and a preference for higher coordination numbers. The similarity of cerium and calcium reflects in their natural occurrence, where cerium is always found together with calcium (e.g., in apatite). In the biological context, the tendency to precipitate together with calcium is reflected in the affinity of cerium to the mineral bone matrix (Jowsey et al. 1958; Ewaldsson and Magnusson 1964; Schmautz 1964) and in the induction of local soft-tissue calcification (Gabbiani et al. 1966; McClure 1980; Garrett and McClure 1981). The mechanism of this calcergic action is not exactly known, but it has been hypothesized that cerium precipitates with pyrophosphate, resulting in the loss of the calcification-inhibitory function of pyrophosphate, and that these precipitates form crystallization nuclei on which calcium and phosphate accumulate to form apatite (Boeckx et al. 1992).

Because of their similar ionic radius, Ce^{3+} and other lanthanide ions are able to replace calcium in many biomolecules, without necessarily substituting for it functionally. For example, they interfere with calcium-dependent reactions involved in the blood clotting cascade such as the activation of prothrombin (Furie et al. 1976) and factor XIII (Achyuthan et al. 1989). This behavior probably accounts for the long-known anticoagulant effects of the lanthanides, which prior to the widespread availability of heparin raised temporary interest in using them as antithrombotic drugs (Vincke and Oelkers1937; Vincke and Schmidt 1942; Beaser et al. 1942; Vincke 1944).

Ce^{3+} is capable of binding to the Ca^{2+}/Mg^{2+}-ATPase of sarcoplasmic reticulum of skeletal muscle (Yamada and Tonomura 1972) and of inhibiting the active transport of calcium through mitochondrial membranes (Mela 1969; Crompton et al. 1979). Moreover, Ce^{3+} is a potent blocker of neuronal low voltage-activated (T-type) calcium channels (Mlinar and Enyeart 1993; Beedle et al. 2002) and of high voltage-activated calcium channels of presynaptic nerve terminals (Nachshen 1984) and skeletal muscle cells (Lansman 1990). Thereby, its interactions with ion channels are not exhausted: A-type potassium channels of adrenal cortical cells are inhibited by binding of Ce^{3+} to sites which are not specific for calcium (Enyeart et al. 1998), whereas currents through type A γ-aminobutyric acid (GABA)-activated chloride channels of rat dorsal root ganglion neurons are enhanced (Ma and Narahashi 1993; Narahashi et al. 1994). Ce^{3+} is also capable of binding to the calcium-binding sites of the N-terminal domain of calmodulin (CaM), which mediates intracellular responses to calcium fluxes, in a cooperative manner (Bentrop et al. 1997) and of substituting for calcium in the regulation of calcium/CaM-dependent enzymes such as phosphorylase kinase (Sotiroudis 1986).

The strong interrelation between these two ions is further illustrated by the fact that cerium(III) chloride administered intravenously to rats induces an increase of serum calcium levels over a certain dose range, with high doses being ineffective (Johansson et al. 1968). The reason for this phenomenon is unknown, but it has to be kept in mind that a variety of biological actions of the lanthanides do not follow simple monotonic dose-response relationships, but may turn from enhancing to inhibiting, depending on the dose (Evans 1990; Wang et al. 2003).

Antiemetic properties

The first documented medical use of cerium dates back to the mid nineteenth century when the obstetrician James Y. Simpson reported on satisfactory therapeutic results obtained with cerium nitrate for the relief of vomiting (Simpson 1854). Especially in vomiting of pregnancy, oral administration of cerium(III) oxalate has been widely practiced during the following decades. Its use has also been proposed for other forms of vomiting such as in cases of sea-sickness, for other gastrointestinal disorders such as chronic diarrhea, and even in neurological disorders such as epilepsy and chorea. The medicinal preparations used until the early twentieth century contained substantial and varying amounts of other lanthanides, however, supposedly without altering its therapeutic effects (Böhm 1915; Wilcox 1916).

As controversial as the opinions regarding the therapeutic value of cerium(III) oxalate were the tentative explanations for its effects, ranging from sedative effects on the cerebral vomiting center or on the pneumogastric nerve and a lowering effect on the reflex excitability of the gastrointestinal tract to astringent properties or the mere formation of a protective coating on the wall of the stomach (Baehr and Wessler 1908; Hara 1923; Umezawa 1925). Objections to this apparently nontoxic drug mainly arose from the facts that it is nearly insoluble in water, except for strongly acidic solutions, and that it is not absorbed from the gastrointestinal tract to any meaningful extent, making pharmacological actions other than locally gastric rather unlikely (Wilcox 1916). The basis for its use was seriously challenged by studies in dogs, which indicated marked effects on vomiting caused by local irritation of the stomach, but either no or only a weak inhibitory effect on vomiting of central nervous origin as induced by apomorphine (Baehr and Wessler 1908; Umezawa 1925). Since some soluble cerium(III) salts are actually capable of preventing apomorphine-induced vomiting when given intravenously, the modest effects of oral cerium(III) oxalate observed in these studies can easily be explained by its poor intestinal absorption (Umezawa 1925).

The low bioavailability and the apparently unreliable therapeutic properties soon led to the development of various soluble cerium preparations with the aim of overcoming these limitations (Böhm 1915). A formulation described as colloidal cerium(III) oxalate (Lange 1933) was available until the 1950s, with vomiting of pregnancy and all forms of motion sickness as primary indications, but it was later blended with (and eventually replaced by) the antihistaminic meclizine (Gigglberger and Höhn 1958). Unfortunately, the efficacy of cerium(III) oxalate has never been objectivized by clinical studies, and interest in elucidating its mode of action ceased after its use had been abandoned.

It has long been known that Ce^{3+} reduces the contractility of the heart (Mines 1910) and skeletal muscles by inhibiting neuromuscular transmission (Hara 1923). It also affects excitation-contraction coupling in intestinal smooth muscle, where it inhibits both the phasic and tonic components of the response to a muscarinic agonist (Triggle and Triggle 1976). Since Ce^{3+} generally interferes with basic muscular and neuronal functions mainly due to its calcium-antagonistic actions, it is tempting to assume a connection between this behavior and its antiemetic properties.

Antiseptic and immunomodulatory properties with special reference to the treatment of burns

The bacteriostatic effects of cerium(III) nitrate and other cerium compounds had already been recognized near the end of the nineteenth century (Drossbach 1897), soon leading to their use as topical antiseptics in human and veterinary medicine. Apart from "dymal," a mixture of rare earth salicylates, several preparations were marketed under the common name "ceolat," namely cerium(III) acetate solutions as well as powders and ointments containing cerium(III) stearate. These remedies have been beneficial in the treatment of wounds, including burns, weeping eczema, intertrigo and decubitus, skin gangrene, impetigo contagiosa, and other skin diseases (Böhm 1915). The idea that the oxidizing properties of cerium(IV) might add to the bacteriostatic effects led to the use of cerium(IV) potassium sulfate, which was applied as an antiseptic powder under the name "ceriform" (Böhm 1915).

Systematic investigations later confirmed the bacteriostatic and bactericidal activity of cerium(III) chloride, cerium(III) nitrate, and cerium(IV) sulfate in a wide variety of bacteria (Burkes and McCleskey 1947) and showed that gram-negative bacteria tend to be somewhat more susceptible than gram-positive bacteria (Muroma 1958). These findings formed the basis for the clinical evaluation of topical cerium(III) nitrate (applied as a cream or in saline solution) in the treatment of extensive, life-threatening burns, with the encouraging result of a nearly 50% reduction in the death rate as compared to the mortality anticipated if the patients had been treated with silver nitrate (Monafo et al. 1976).

Because the flora recovered from treated wounds tended to be dominated by gram-positive bacteria, combination with the complementary acting silver sulfadiazine has been recommended (Monafo et al. 1976). Death from sepsis in patients surviving the original hemodynamic and pulmonary insults from near-total burns could be effectively prevented with the use of this combination (Fox et al. 1977; Monafo et al. 1977, 1978), and halving of the mortality rate as compared to prediction was confirmed (Monafo 1983). Although results of in vitro studies concerning the synergism or antagonism of these drugs have been conflicting (Rosenkranz 1979; Heggers et al. 1979; Saffer et al. 1980; Holder 1982; Marone et al. 1998) and the first prospective, randomized studies comparing silver sulfadiazine plus cerium nitrate to silver sulfadiazine alone failed to demonstrate any advantage of the combination in adults (Munster et al. 1980) and children (Bowser et al. 1981), a more recent trial confirmed the greater efficacy of the combination in terms of faster re-epithelialization, earlier readiness for autologous skin grafting, and reduced duration of hospitalization and even suggests a reduced mortality (de Gracia 2001).

Further beneficial aspects of the cerium nitrate adjunct have been emphasized: a reduced number of highly colonized wound sites (Hermans 1984) and the formation of a dry, adherent eschar which allows an excellent take of graft after excision (Ross et al. 1993) and postponing of skin grafting (Wassermann et al. 1989; Hadjiiski and Lesseva 1999). The characteristic leathery consistency of these eschars, which form upon contact of dermal collagen with cerium, at least in part results from superficial calcification induced by the calcergic action of cerium and constitutes an additional, physical barrier that insulates the dermis from germ contamination (Boeckx et al. 1992). Penetration through the eschar and skin is low (Boeckx et al. 1992; Herruzo-Cabrera et al. 1992), and no cerium could be detected in blood and urine even after treatment of large wounds for several weeks (Monafo et al. 1976).

Apart from direct antibacterial effects, immunomodulatory properties have been recognized as a major mechanism by which cerium helps prevent sepsis in burn patients. In mice, cerium nitrate protects from postburn immunosuppression, as indicated by the preservation of hypersensitivity reactions following antigen challenge of previously sensitized animals (Hansbrough et al. 1984; Peterson et al. 1985), reduced alterations of the splenic helper to suppressor lymphocyte ratio (Zapata-Sirvent and Hansbrough 1985), and improved survival following septic challenge (Zapata-Sirvent et al. 1986). In humans, it seems to preserve normal T-cell functions such as the production of interleukin 2 (IL-2) and IL-2 receptor expression (Sparkes 1993; Allgöwer et al. 1995). This effect probably results from the neutralizing action on the immunosuppressive burn toxins formed at the wound site. In particular, cerium irreversibly denaturizes the toxic lipid-protein complex (LPC), thereby probably fixing it in the eschar and preventing its entry into the circulation (Allgöwer et al. 1995). Additionally, it inhibits the calcium-dependent hemolytic activity of a low molecular weight lipid-peptide complex termed SAP isolated from the serum of burn patients (Ninnemann et al. 1985). A study in mice receiving skin autografts, which were burned and treated or not with cerium nitrate in vitro, demonstrated the protective effect against the mortality induced by grafting burned skin. Because the observed survival rate was nearly identical to that of burned animals after necrotomy and subsequent skin grafting, the effect of cerium nitrate was considered a chemical equivalent to surgical wound excision and recommended in cases not allowing early surgical intervention (Kistler et al. 1990). An impressive reduction of the anticipated death rate following treatment of patients with a single early bathing in aqueous cerium nitrate solution supports this opinion (Scheidegger et al. 1992).

Since it has been recognized that, apart from sepsis, a systemic inflammatory response characterized by a chaotic cytokine cascade, which is responsible for an eventual multiple organ dysfunction syndrome, is another major cause of "late" mortality in burn patients, new aspects of the neutralizing action of cerium on LPC have become evident (Allgöwer et al. 1995; Sparkes 1997). Recent experimental studies focus on the protective effect against the processes involved in the systemic inflammatory response (Deveci et al. 2000; Eski et al. 2001).[1]

Apart from these indirect effects specific for the burn setting, cerium also possesses direct immunomodulatory properties. Injections of cerium(III) salts in rats inhibit edematous inflammation caused by inflammatory agents (Jancsó and Jancsó-Gábor 1960) and inhibit reticuloendothelial system functions, in particular the phagocytotic activity of Kupffer cells (Lazar 1973). Cerium(III) chloride is capable of interfering with epidermal Langerhans cell functions by inhibiting their Ca^{2+}/Mg^{2+}-dependent ATPase (Gruner et al. 1991) and of inhibiting histamine release from basophil granulocytes (Gruner et al. 1992). Because of the pathogenetic role of both histamine-releasing mast cells and Langerhans cells in atopic eczema, therapeutic use of lanthanides has been proposed.

[1] The cerium nitrate–silver sulfadiazine cream marketed under the name "flammacerium" in several European countries and much advocated by clinicians (Lansdown et al. 2003), but classified as an orphan drug in the USA, was subject to an accelerated approval prior to the termination of a clinical phase III trial to respond to the urgent need resulting from the terror attacks of Sept. 11, 2001.

Antineoplastic properties

Dating back to the pioneering era of cancer chemotherapy, two clinical reports on the use of "introcid," i.e., a solution of cerium(III) iodide for intravenous administration, in the treatment of patients with lymphogranulomatosis (M. Hodgkin) or inoperable solid tumors deserve mentioning. This experimental drug has been explored for its antineoplastic properties among other compounds of iodine which were expected to accumulate in tumor tissue. According to the clinical reports, cerium(III) iodide has been applied with remarkable benefit, as indicated by tumor shrinkage and improved quality of life in several cases of locally advanced or metastatic tumors of different origin (Lewin 1924; Cohn 1925). The authors judged the therapeutic effects as superior to those of other compounds of iodine and therefore considered cerium either the active component or an activator for iodine. Animal experiments using Kato's rabbit sarcoma confirmed the superiority to other compounds of iodine, in particular potassium iodide and calcium iodide which lack any appreciable activity in this tumor model (Ito 1937).

These observations are in peculiar contrast to the poor results obtained with cerium(III) chloride in experimental tumor models. While this compound is completely devoid of activity in two transplantable rat tumors (Maxwell and Bischoff 1931), it exerts moderate inhibitory effects on the proliferation of malignant cells in vitro, requiring high micromolar or even millimolar concentrations. The antiproliferative effects have been reported to be associated with morphological changes (polyhedral spreading) and cell cycle arrest in the G_0/G_1 phase in melanoma cells (Sato et al. 1998), decreased calmodulin levels in leukemia cells, upregulation of p53 and p21 tumor suppressor gene expression in gastric cancer cells (Ji et al. 2000), and apoptosis in leukemia cells (Dai et al. 2002).

Cerium(III) complexes of coumarin derivatives with known antibacterial, anticoagulant, and cytotoxic properties have been synthesized and explored for their cytotoxicity in malignant cells. In particular, $Ce(HL)_3 \cdot 5\ H_2O$ with L=4-hydroxy-3-[1-(4-chlorophenyl)-3-oxobutyl]-2H-1-benzopyran-2-one (coumachlor) (Kostova et al. 1999), $Ce(HL)_3 \cdot 4\ H_2O$ with L=4-hydroxy-3-[1-(4-nitrophenyl)-3-oxobutyl]-2H-1-benzopyran-2-one (niffcoumar) (Manolov et al. 1999), and $Ce(HL)_2(OH) \cdot 5\ H_2O$ with L=4-methyl-7-hydroxy-2H-1-benzopyran-2-one (mendiaxon) (Kostova et al. 2001) exert moderate cytotoxic effects in lymphoma and leukemia cells, while $Ce(HL)_3 \cdot 4\ H_2O$ with L=4-hydroxy-3-(3-oxo-1-phenylbutyl)-2H-1-benzopyran-2-one (warfarin) is devoid of cytoxicity in relevant concentrations (Kostova et al. 1999).

Cerium(III) complexes with 2,2′-bipyridine, 1,10-phenthroline and related ligands have been synthesized and evaluated in vitro. Among these complexes, those with 1,10-phenanthroline appear most promising. In particular, *trans*-[aquachlorobis(1,10-phenanthroline)cerium(III)] dichloride (KP776) exerts strong cytotoxic effects in cancer cells, with IC_{50} values mainly in the low micromolar range. Experiments in cell models, each comprised of a parental cell line and a doxorubicin-selected subline overexpressing either P-glycoprotein or MRP1, revealed that KP776 is substrate to none of these two multidrug resistance-conferring transport proteins, but that the multidrug-resistant sublines show an unexpected collateral sensitivity to this compound (Jakupec et al. 2002). Moreover, all attempts to induce resistance to KP776 in cancer cell lines by prolonged exposure to minimally cytotoxic concentrations of the compound failed (P. Heffeter and W. Berger, personal communication). KP776 has a much lower DNA interstrand cross-linking efficiency than cisplatin and induces neither strand breaks nor alterations of the secondary structure of plasmids (Jakupec et al. 2002). Thus, the mechanisms by which KP776 and related compounds exert their effects on malignant cells remain to be elucidated. Given the pivotal role of calcium signaling in controlling cell cycle and proliferation (Short et al. 1993; Takuwa et al. 1995; Munaron et al. 2004) as well as malignant progression and angiogenesis (Kamrava et al. 2002), it seems reasonable to hypothesize a connection between the capability of cerium to interfere with calcium-dependent processes and its antineoplastic potential.

References

Achyuthan KE, Mary A, Greenberg CS (1989) Tb(III)-ion-binding-induced conformational changes in platelet factor XIII. Biochem J 257:331–338

Allgöwer M, Schoenenberger GA, Sparks BG (1995) Burning the largest immune organ. Burns 21(1):S7–S47

Asprey LB, Cunningham BB (1960) Unusual oxidation states of some actinide and lanthanide elements. Progr Inorg Chem 2:267–302

Baehr G, Wessler H (1908) The use of cerium oxalate for the relief of vomiting: an experimental study of the effects of some salts of cerium, lanthanum, neodymium and thorium. Arch Intern Med 2:517–531

Beaser SB, Segel A, Vandam L (1942) The anticoagulant effects in rabbits and man of the intravenous injection of salts of the rare earths. J Clin Invest 21:447–454

Beedle AM, Hamid J, Zamponi GW (2002) Inhibition of transiently expressed low- and high-voltage-activated calcium channels by trivalent metal cations. J Membr Biol 187:225–238

Bentrop D, Bertini I, Cremonini MA, Forsén S, Luchinat C, Malmendal A (1997) Solution structure of the paramagnetic complex of the N-terminal domain of calmodulin with two Ce^{3+} ions by ^{1}H NMR. Biochemistry 36:11605–11618

Boeckx W, Blondeel PN, Vandersteen K, De Wolf-Peters C, Schmitz A (1992) Effect of cerium nitrate–silver sulphadiazine on deep dermal burns: a histological hypothesis. Burns 18:456–462

Böhm CR (1915) Die seltenen Erden in der Therapie. Z Angew Chem 28:333–336, 346–348

Bowser BH, Caldwell FT Jr, Cone JB, Eisenach KD, Thompson CH (1981) A prospective analysis of silver sulfadiazine with and without cerium nitrate as a topical agent in the treatment of severely burned children. J Trauma 21:558–563

Burkes S, McCleskey CS (1947) The bacteriostatic activity of cerium, lanthanum, and thallium. J Bacteriol 54:417–424

Cohn B (1925) Erfahrungen mit Introcid bei der Palliativbehandlung inoperabler Karzinome. Dtsch Med Wochenschr 51:1984–1986

Crompton M, Heid I, Baschera C, Carafoli E (1979) The resolution of calcium fluxes in heart and liver mitochondria using the lanthanide series. FEBS Lett 104:352–354

Dai Y, Li J, Li J, Yu L, Dai G, Hu A, Yuan L, Wen Z (2002) Effects of rare earth compounds on growth and apoptosis of leukemic cell lines. In Vitro Cell Dev Biol Anim 38:373–375

de Gracia CG (2001) An open study comparing topical silver sulfadiazine and topical silver sulfadiazine–cerium nitrate in the treatment of moderate and severe burns. Burns 27:67–74

Deveci M, Eski M, Sengezer M, Kisa U (2000) Effects of cerium nitrate bathing and prompt burn wound excision on IL-6 and TNF-α levels in burned rats. Burns 26:41–45

Drossbach GP (1897) Ueber den Einfluss der Elemente der Cer- und Zircongruppe auf das Wachstum von Bakterien. Zentralbl Bakteriol Parasitenk Infektionskrankh Abt 1, 21:57–58

Enyeart JJ, Xu L, Gomora JC, Enyeart JA (1998) Modulation of I_A potassium current in adrenal cortical cells by a series of ten lanthanide elements. J Membr Biol 164:139–153

Eski M, Deveci M, Celiköz B, Nisanci M, Türegün M (2001) Treatment with cerium nitrate bathing modulate systemic leukocyte activation following burn injury: an experimental study in rat cremaster muscle flap. Burns 27:739–746

Evans CH (1990) Biochemistry of the lanthanides. Biochemistry of the elements 8. Plenum Press, New York

Ewaldsson B, Magnusson G (1964) Distribution of radiocerium and radiopromethium in mice. An autoradiographic study. Acta Radiol Ther Phys Biol 2:65–72

Eyring LR (1964) Progress in the science and technology of the rare earths. Pergamon Press, London

Fox CL Jr, Monafo WW Jr, Ayvazian VH, Skinner AM, Modak S, Stanford J, Condict C (1977) Topical chemotherapy for burns using cerium salts and silver sulfadiazine. Surg Gynecol Obstet 144:668–672

Furie BC, Mann KG, Furie B (1976) Substitution of lanthanide ions for calcium ions in the activation of bovine prothrombin by activated factor X. High affinity metal-binding sites of prothrombin and the derivatives of prothrombin activation. J Biol Chem 251:3235–3241

Gabbiani G, Jacqmin ML, Richard RM (1966) Soft-tissue calcification induced by rare earth metals and its prevention by sodium pyrophosphate. Br J Pharmacol Chemother 27:1–9

Garrett IR, McClure J (1981) Lanthanide-induced calcergy. J Pathol 135:267–275

Gigglberger H, Höhn H (1958) Ein Beitrag zur Prüfung des Antiemeticums Peremesin-forte. Med Klin 53:226–227

Gruner S, Diezel W, Strunk D, Eckert R, Siems W, Anhalt GJ (1991) Inhibition of Langerhans cell ATPase and contact sensitization by lanthanides–role of T-suppressor cells. J Invest Dermatol 97:478–482

Gruner S, Sehrt I, Müller GM, Zwirner A, Strunk D, Sönnichsen N (1992) Inhibition of histamine release from human granulocytes by ions of the rare earth elements lanthanum and cerium. Agents Actions 36:207–211

Hadjiiski OG, Lesseva MI (1999) Comparison of four drugs for local treatment of burn wounds. Eur J Emerg Med 6:41–47

Hansbrough JF, Zapata-Sirvent R, Peterson V, Wang X, Bender E, Claman H, Boswick J (1984) Characterization of the immunosuppressive effect of burned tissue in an animal model. J Surg Res 37:383–393

Hara S (1923) Beiträge zur Pharmakologie der seltenen Erdmetalle. I. Mitteilung: Über das Cerium. Arch Exp Pathol Pharmakol 100:217–253

Heggers JP, Ko F, Robson MC (1979) Cerium nitrate/silver sulphadiazine: synergism or antagonism as determined by minimum inhibitory concentration. Burns 5:308–311

Hermans RP (1984) Topical treatment of serious infections with special reference to the use of a mixture of silver sulphadiazine and cerium nitrate: two clinical studies. Burns 11:59–62

Herruzo-Cabrera R, Garcia-Torres V, Rey-Calero J, Vizcaino-Alcaide MJ (1992) Evaluation of the penetration strength, bactericidal efficacy and spectrum of action of several antimicrobial creams against isolated microorganisms in a burn centre. Burns 18:39–44

Holder IA (1982) In vitro inactivation of silver sulphadiazine by the addition of cerium salts. Burns 8:274–277

Ito H (1937) An experimental study on the effect of iodides on the growth and radiosensitivity of malignant tumour. Part I. The effect of iodides on the growth of malignant tumour. Jpn J Obstet Gynecol 20:536–540

Jakupec MA, Arion V, Berger W, Wild S, Zorbas H, Keppler BK (2002) Novel antineoplastic complexes of bismuth(III), cerium(III) and lanthanum(III). Eur J Cancer 38 [suppl 7]:S125

Jancsó N, Jancsó-Gábor A (1960) Wirkung von Antikoagulantien auf entzündliche Gewebsreaktionen. Naunyn-Schmiedebergs Arch Exp Pathol Pharmakol 238:83–84

Ji Y-J, Xiao B, Wang ZH, Cui MZ, Lu YY (2000) The suppression effect of light rare earth elements on proliferation of two cancer cell lines. Biomed Environ Sci 13:287–292

Johansson O, Perrault G, Savoie L, Tuchweber B (1968) Action of various metallic chlorides on calcaemia and phosphataemia. Br J Pharmacol Chemother 33:91–97

Jowsey J, Rowland RE, Marshall JH (1958) The deposition of the rare earths in bone. Radiat Res 8:490–501

Kamrava MR, Michener CM, Kohn EC (2002) Calcium signaling as a molecular target in cancer. Pharm News 9:435–442

Kistler D, Hafemann B, Schoenenberger GA, Hettich R (1990) Increased survival rates by topical treatment of burns with cerium nitrate. Eur Surg Res 22:283–290

Kleber EV (1961) Selten-Erden Forschung. MacMillan, New York

Komiyama M, Sumaoka J (2001) Complex formation of sugars and lanthanide ions and its applications. Kikan Kagaku Sosetsu 48:129–136

Kostova I, Manolov I, Konstantinov S, Karaivanova M (1999) Synthesis, physicochemical characterization and cytotoxic screening of new complexes of cerium, lanthanum and neodymium with Warfarin and Coumachlor sodium salts. Eur J Med Chem 34:63–68

Kostova I, Manolov I, Nicolova I, Konstantinov S, Karaivanova M (2001) New lanthanide complexes of 4-methyl-7-hydroxycoumarin and their pharmacological activity. Eur J Med Chem 36:339–347

Lange J (1933) Peremesin, ein giftfreies Mittel gegen Erbrechen (besonders Schwangerschaftserbrechen, See-, Luft-, Eisenbahnkrankheit etc.). Münch Med Wochenschr 80:1327–1329

Lansdown ABG, Myers SR, Clarke JA, O'Sullivan P (2003) A reappraisal of the role of cerium in burn wound management. J Wound Care 12:113–118

Lansman JB (1990) Blockade of current through single calcium channels by trivalent lanthanide cations. Effect of ionic radius on the rates of ion entry and exit. J Gen Physiol 95:679–696

Lazar G (1973) Reticuloendothelial-blocking effect of rare earth metals in rats. J Reticuloendothel Soc 13:231–237

Lewin C (1924) Über die Verwendung einer Cerium-Jodverbindung (Introcid) in der Therapie der Geschwulstbildungen. Med Klin 20:1319–1323

Ma JY, Narahashi T (1993) Enhancement of γ-aminobutyric acid–activated chloride channel currents by lanthanides in rat dorsal root ganglion neurons. J Neurosci 13:4872–4879

Manolov I, Kostova I, Konstantinov S, Karaivanova M (1999) Synthesis, physicochemical characterization and cytotoxic screening of new complexes of cerium, lanthanum and neodymium with Niffcoumar sodium salt. Eur J Med Chem 34:853–858

Marone P, Monzillo V, Perversi L, Carretto E (1998) Comparative in vitro activity of silver sulfadiazine, alone and in combination with cerium nitrate, against staphylococci and gram-negative bacteria. J Chemother 10:17–21

Maxwell LC, Bischoff F (1931) Studies in cancer chemotherapy. X. The effect of thorium, cerium, erbium, yttrium, didymium, praseodymium, manganese, and lead upon transplantable rat tumors. J Pharmacol Exp Ther 43:61–70

McClure J (1980) The production of heterotopic calcification by certain chemical salts. J Pathol 131:21–33

Mela L (1969) Inhibition and activation of calcium transport in mitochondria. Effect of lanthanides and local anesthetic drugs. Biochemistry 8:2481–2486

Mines GR (1910) The action of beryllium, lanthanum, yttrium and cerium on the frog's heart. J Physiol 40:327–346

Mlinar B, Enyeart JJ (1993) Block of current through T-type calcium channels by trivalent metal cations and nickel in neural rat and human cells. J Physiol 469:639–652

Moeller T (1963) Die Chemie der Lanthaniden. Reinhold, New York

Moeller T (1973) Lanthanides. In: Bailar JC Jr, Emeleus HJ, Nyholm R, and Trotman-Dickeson AT (eds) Comprehensive inorganic chemistry, vol. 4. Pergamon Press, Oxford, pp 1–101

Monafo L (1983) The use of topical cerium nitrate-silver sulfadiazine in major burn injuries. Panminerva Med 25:151–154

Monafo WW, Tandon SN, Ayvazian VH, Tuchschmidt J, Skinner AM, Deitz F (1976) Cerium nitrate: a new topical antiseptic for extensive burns. Surgery 80:465–473

Monafo WW, Ayvazian VH, Skinner AM (1977) Control of infection in major burn wounds by cerium nitrate/silver sulphadiazine. Burns 3:104–107

Monafo WW, Robinson HN, Yoshioka T, Ayvazian VH (1978) "Lethal" burns. Arch Surg 113:397–401

Munaron L, Antoniotti S, Pla AF, Lovisolo D (2004) Blocking Ca^{2+} entry: a way to control cell proliferation. Curr Med Chem 11:1533–1543

Munster AM, Helvig E, Rowland S (1980) Cerium nitrate–silver sulfadiazine cream in the treatment of burns: a prospective evaluation. Surgery 88:658–660

Muroma A (1958) Bactericidal action of salts of certain rare earth metals. Ann Med Exp Biol Fenn 36 [suppl 6]:1–54

Nachshen DA (1984) Selectivity of the Ca binding site in synaptosome Ca channels. Inhibition of Ca influx by multivalent metal cations. J Gen Physiol 83:941–967

Narahashi T, Ma JY, Arakawa O, Reuveny E, Nakahiro M (1994) GABA receptor-channel complex as a target site of mercury, copper, zinc, and lanthanides. Cell Mol Neurobiol 14:599–621

Nash KL, Sullivan JC (1991) Kinetics of complexation and redox reactions of the lanthanoids in aqueous solutions. In: Gschneidner Jr KA and Eyring L (eds) Handbook on the physics and chemistry of the rare earths, vol 15. Elsevier, Amsterdam, pp 287–347

Ninnemann JL, Ozkan AN, Sullivan JJ (1985) Hemolysis and suppression of neutrophil chemotaxis by a low molecular weight component of human burn patient sera. Immunol Lett 10:63–69

Peterson VM, Hansbrough JF, Wang XW, Zapata-Sirvent R, Boswick JA (1985) Topical cerium nitrate prevents postburn immunosuppression. J Trauma 25:1039–1044

Rosenkranz HS (1979) A synergistic effect between cerium nitrate and silver sulphadiazine. Burns 5:278–281

Ross DA, Phipps AJ, Clarke JA (1993) The use of cerium nitrate–silver sulphadiazine as a topical burns dressing. Br J Plast Surg 46:582–584

Saffer LD, Rodeheaver GT, Hiebert JM, Edlich RF (1980) In vivo and in vitro antimicrobial activity of silver sulfadiazine and cerium nitrate. Surg Gynecol Obstet 151:232–236

Sato T, Hashizume M, Hotta Y, Okahata Y (1998) Morphology and proliferation of B16 melanoma cells in the presence of lanthanoid and Al^{3+} ions. BioMetals 11:107–112

Scheidegger D, Sparkes BG, Lüscher N, Schoenenberger GA, Allgöwer M (1992) Survival in major burn injuries treated by one bathing in cerium nitrate. Burns 18:296–300

Schmautz E (1964) Geschlechtabhängigkeit der Verteilung von Radiocer in der Ratte. Strahlenther 123:267–278

Shigekawa H, Ishida M, Miyake K, Shioda R, Iijima Y, Imai T, Takahashi H, Sumaoka J, Komiyama M (1999) Extended X-ray absorption fine structure study on the cerium(IV)-induced DNA hydrolysis: implication to the roles of 4f orbitals in the catalysis. Appl Phys Lett 74:460–462

Short AD, Bian J, Ghosh TK, Waldron RT, Rybak SL, Gill DL (1993) Intracellular Ca^{2+} pool content is linked to control of cell growth. Proc Natl Acad Sci USA 90:4986–4990

Simpson JY (1854) Note on the therapeutic action of the salts of cerium. Month J Med Sci 19:564

Sinha SP (1966) Complexes of the rare earths. Pergamon Press, New York

Sotiroudis TG (1986) Lanthanide ions and Cd^{2+} are able to substitute for Ca^{2+} in regulating phosphorylase kinase. Biochem Int 13:59–64

Sparkes BG (1993) Mechanisms of immune failure in burn injury. Vaccine 11:504–510

Sparkes BG (1997) Immunological responses to thermal injury. Burns 23:106–113

Sumaoka J, Azuma Y, Komiyama M (1998) Enzymic manipulation of the fragments obtained by cerium(IV)-induced DNA scission: characterization of hydrolytic termini. Chem Eur J 4:205–209

Takarada T, Yashiro M, Komiyama M (2000) Catalytic hydrolysis of peptides by cerium(IV). Chem Eur J 6:3906–3913

Takuwa N, Zhou W, Takuwa Y (1995) Calcium, calmodulin and cell cycle progression. Cell Signal 7:93–104

Trifonov DN (1963) The rare-earth elements. MacMillan, New York

Triggle CR, Triggle DJ (1976) An analysis of the action of cations of the lanthanide series on the mechanical responses of guinea-pig ileal longitudinal muscle. J Physiol 254:39–54

Umezawa R (1925) Über die Wirkung der Ceriumsalze auf das Brechzentrum. Z Gesamte Exp Med 44:404–417

Vincke E (1944) Untersuchungen über Thromboseprophylaxe mit organischen Salzen seltener Erden. I. Mitteilung. Darstellung der Salze und Wirkung auf das Blut. Z Gesamte Exp Med 113:522–535

Vincke E, Oelkers HA (1937) Zur Pharmakologie der seltenen Erden. I. Mitteilung: Wirkung auf die Blutgerinnung. Naunyn-Schmiedebergs Arch Exp Pathol Pharmakol 187:594–603

Vincke E, Schmidt E (1942) Über den Prothrombinmangel durch seltene Erden und seine Beeinflussung durch Vitamin K_5. Hoppe-Seyler's Z Physiol Chem 273:39–46

Wang K, Cheng Y, Yang X, Li R (2003) Cell responses to lanthanides and potential pharmacological actions of lanthanides. In: Sigel A and Sigel H (eds) Metal ions in biological systems, vol 40. Marcel Dekker, New York, 707–751

Wassermann D, Schlotterer M, Lebreton F, Levy J, Guelfi MC (1989) Use of topically applied silver sulphadiazine plus cerium nitrate in major burns. Burns 15:257–260

Wilcox RW (1916) The therapeutics of cerium. N Y Med J 114:836–838

Yamada S, Tonomura Y (1972) Reaction mechanism of Ca^{2+}-dependent ATPase of sarcoplasmic reticulum from skeletal muscle. VII. Recognition and release of Ca^{2+} ions. J Biochem 72:417–425

Zapata-Sirvent RL, Hansbrough JF (1985) Postburn immunosuppression in an animal model. III: Maintenance of normal splenic helper and suppressor lymphocyte subpopulations by immunomodulating drugs. Surgery 97:721–727

Zapata-Sirvent RL, Hansbrough JF, Bender EM, Bartle EJ, Mansour MA, Carter WH (1986) Postburn immunosuppression in an animal model. IV. Improved resistance to septic challenge with immunomodulating drugs. Surgery 99:53–59

Printing: Krips bv, Meppel
Binding: Litges & Dopf, Heppenheim